500

AP Chemistry Questions

to know by test day

Get ready for your AP exam with McGraw-Hill's ***5 STEPS TO A 5: 500 AP Chemistry Questions to Know by Test Day***!

Also in the 5 Steps series:

5 Steps to a 5: AP Chemistry

Also in the 500 AP Questions to Know by Test Day series:

5 Steps to a 5: 500 AP Biology Questions to Know by Test Day
5 Steps to a 5: 500 AP Calculus Questions to Know by Test Day
5 Steps to a 5: 500 AP English Language Questions to Know by Test Day
5 Steps to a 5: 500 AP English Literature Questions to Know by Test Day
5 Steps to a 5: 500 AP Environmental Science Questions to Know by Test Day
5 Steps to a 5: 500 AP European History Questions to Know by Test Day
5 Steps to a 5: 500 AP Human Geography Questions to Know by Test Day
5 Steps to a 5: 500 AP Microeconomics/Macroeconomics Questions to Know by Test Day
5 Steps to a 5: 500 AP Physics Questions to Know by Test Day
5 Steps to a 5: 500 AP Psychology Questions to Know by Test Day
5 Steps to a 5: 500 AP Statistics Questions to Know by Test Day
5 Steps to a 5: 500 AP U.S. Government & Politics Questions to Know by Test Day
5 Steps to a 5: 500 AP U.S. History Questions to Know by Test Day
5 Steps to a 5: 500 AP World History Questions to Know by Test Day

500 AP Chemistry Questions to know by test day

Mina Lebitz

New York Chicago San Francisco Lisbon London Madrid Mexico City
Milan New Delhi San Juan Seoul Singapore Sydney Toronto

The McGraw-Hill Companies

1 2 3 4 5 6 7 8 9 10 FGR/FGR 1 9 8 7 6 5 4 3 2 1

ISBN 978-0-07-177405-5
MHID 0-07-177405-X

e-ISBN 978-0-07-177406-2
e-MHID 0-07-177406-8

Library of Congress Control Number: 2011936615

Series interior design by Jane Tenenbaum.

This book is printed on acid-free paper.

CONTENTS

ABOUT THE AUTHOR

Mina Lebitz has a BS in biology from the State University of New York at Albany and an MS in nutritional biochemistry from Rutgers University. She has more than 16 years of teaching experience at both the high school and college level. Ms. Lebitz received the *New York Times*' Teachers Who Make a Difference award in 2003 during her tenure at Brooklyn Technical High School and was the senior science tutor at one of the most prestigious tutoring and test prep agencies in the United States. Currently, she is doing research, writing, and assisting students in reaching their academic goals, while continuing to learn everything she can about science. Her website is www.idigdarwin.com.

INTRODUCTION

Congratulations! You've taken a big step toward AP success by purchasing *5 Steps to a 5: 500 AP Chemistry Questions to Know by Test Day*. We are here to help you take the next step and earn a high score on your AP Exam so you can earn college credits and get into the college or university of your choice.

This book gives you 500 AP-style multiple-choice questions that cover all the most essential course material. Each question has a detailed answer explanation. These questions will give you valuable independent practice to supplement both your regular textbook and the groundwork you are already covering in your AP classroom. This and the other books in this series were written by expert AP teachers who know your exam inside and out and can identify crucial exam information and questions that are most likely to appear on the test.

You might be the kind of student who takes several AP courses and needs to study extra questions a few weeks before the exam for a final review. Or you might be the kind of student who puts off preparing until the last weeks before the exam. No matter what your preparation style is, you will surely benefit from reviewing these 500 questions that closely parallel the content, format, and degree of difficulty of the questions on the actual AP exam. These questions and their answer explanations are the ideal last-minute study tool for those final few weeks before the test.

Remember the old saying "Practice makes perfect." If you practice with all the questions and answers in this book, we are certain you will build the skills and confidence needed to do great on the exam. Good luck!

—Editors of McGraw-Hill Education

NOTE FROM THE AUTHOR

The AP Chemistry exam has a multiple-choice section during which you are not allowed to use a calculator. In this section, you can round fairly generously to solve problems faster. However, do not round generously in the free-response problems. Calculators are allowed for most of the free-response section so you're expected to be precise in your calculations (and obey the rules for significant figures).

The questions in this book cover the material for both the free-response and multiple-choice sections of the AP Chemistry exam. Some calculations are best done with a calculator, but all of them can be done without one. In the multiple-choice section of the AP Chemistry exam, there will never be a problem that requires the correct answer from a different question to solve. To cover all the material related to the exam in this book, topics from free-response questions have been adapted into multiple-choice style questions. This required that some questions occur in groups referring to a common experiment or data set. These questions may rely on answers from other questions.

500 AP Chemistry Questions to know by test day

Atomic Theory and Structure

1. Which of the following shows the correct number of protons, neutrons, and electrons in a neutral cadmium-112 atom?

	Protons	Neutrons	Electrons
(A)	48	48	48
(B)	48	64	48
(C)	48	64	64
(D)	64	48	64
(E)	112	48	112

Questions 2–7 refer to the following diagram of the periodic table.

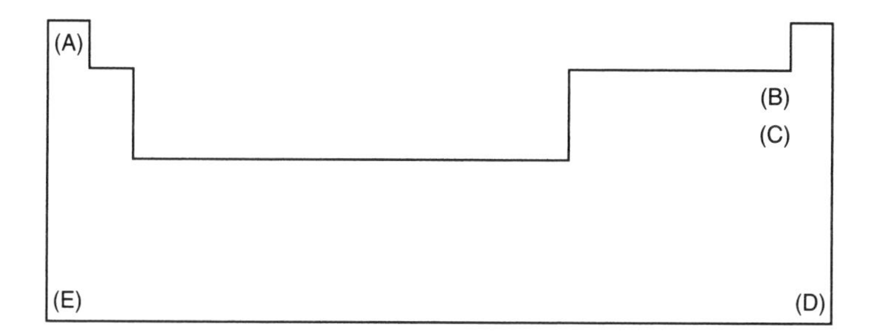

2. Reacts violently with water at 298 K

3. Highest first ionization energy

4. Highest electronegativity

5. Highest electron affinity

6. Largest atomic radius

7. Most metallic character

8. The atomic mass of bromine is 79.904. Given that the only two naturally occurring isotopes are ^{79}Br and ^{81}Br, the abundance of ^{79}Br isotope is approximately:
(A) 20 percent
(B) 40 percent
(C) 50 percent
(D) 80 percent
(E) 99 percent

9. The atomic mass of Sr is 87.62. Given that there are only three naturally occurring isotopes of strontium, ^{86}Sr, ^{87}Sr, and ^{88}Sr, which of the following must be true?
(A) ^{86}Sr is the most abundant isotope.
(B) ^{87}Sr is the most abundant isotope.
(C) ^{88}Sr is the most abundant isotope.
(D) ^{86}Sr is the least abundant isotope.
(E) The isotopes ^{87}Sr and ^{88}Sr occur in approximately equal amounts.

10. Which of the following properties generally *decreases* from left to right across a period (from potassium to bromine)?
(A) Electronegativity
(B) Electron affinity
(C) Atomic number
(D) Atomic radius
(E) Maximum value of oxidation number

11. All of the following statements describe the elements of the group 1 alkali metals (not including hydrogen) *except*:
(A) Their reactivity increases with increasing period number.
(B) They have low first ionization energies.
(C) They react violently with water to form strong acids.
(D) They have strong metallic character.
(E) They are all silver solids at 1 atm and 298 K.

12. Which of the following elements would be expected to have chemical properties most similar to those of phosphorus?

(A) S
(B) Se
(C) O
(D) As
(E) Si

13. Which of the following pairs are isoelectronic (have the same number of electrons)?

(A) Kr^-, Br^+
(B) F^-, Na^+
(C) Sc, Ti^-
(D) Be^{2+}, Ne
(E) Cs, Ba^{2+}

14. Which of the following ions has the same number of electrons as I^-?

(A) Sr^{2+}
(B) Rb^+
(C) Cs^+
(D) Ba^{2+}
(E) Br^-

15. Which of the following best explains why the F^- ion is smaller than the O^{2-} ion?

(A) F^- has a more massive nucleus than O^{2-}.
(B) F^- has a higher electronegativity than O^{2-}.
(C) F^- has a greater nuclear charge than O^{2-}.
(D) F^- has a greater number of electrons than O^{2-}.
(E) F^- has more nucleons and electrons than O^{2-}.

16. All of the following are true statements about the periodic table *except*:

(A) The reactivity of the group 1 alkali metals increases with increasing period.
(B) The reactivity of the group 17 halogens decreases with increasing period.
(C) The group 1 and 2 metals react with water to form basic solutions.
(D) The group 18 noble gases can exist only as inert, monatomic gases.
(E) All elements with an atomic number equal to or greater than 84 are radioactive.

17. Which of the following lists contains *all* the diatomic, elemental gases at standard temperatures and pressures?

(A) H, N, O
(B) H, N, O, F, Cl
(C) H, N, O, F, Cl, Br, I
(D) H, N, O, Cl, Br, I, Hg, Rn
(E) H, N, O, Cl, He, Ne, Ar, Kr, Xe, Rn

18. As atomic number increases from 11 to 17 in the periodic table, what happens to atomic radius?

(A) It remains constant.
(B) It increases only.
(C) It decreases only.
(D) It increases, then decreases.
(E) It decreases, then increases.

19. The effective nuclear charge experienced by a valence Kr is different than the effective nuclear charge experienced by a valence electron of K. Which of the following accurately illustrates this difference?

(A) K is a solid while Kr is a gas.
(B) The valence electrons of Kr have a lower first ionization energy than K.
(C) The proton-to-electron ratio is higher for Kr than for K.
(D) Kr has a higher first ionization energy than K.
(E) The valence electrons of Kr experience less shielding by the inner electrons than the valence electrons of K.

Ionization Energies for Element X ($kJ \cdot mol^{-1}$)

First	Second	Third	Fourth	Fifth
786	1,577	3,228	4,354	16,100

20. Based on the ionization energies for element X listed in the table above, which of the following elements is X most likely to be?

(A) Li
(B) Be
(C) Al
(D) Si
(E) As

Questions 21 and 22 refer to the following graph of first ionization energies.

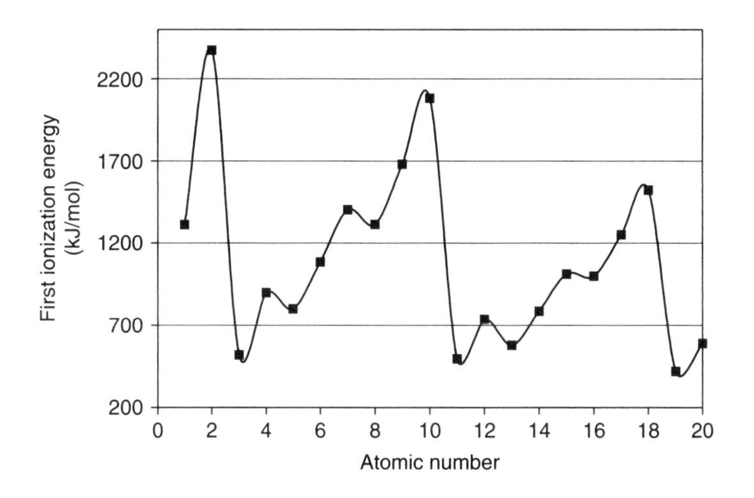

21. Correct explanations for the large drops in ionization energies between elements of atomic numbers 2 and 3, 10 and 11, and 18 and 19 occurs because, compared to elements 3, 11, and 19, elements 2, 10, and 18 have

I. smaller atomic radii.
II. a greater electron affinity.
III. a greater effective nuclear charge.

(A) I only
(B) II only
(C) III only
(D) I and III only
(E) I, II, and III

22. Correct explanations for the increases and decreases in ionization energies between elements between atomic numbers 2 and 10 (and 11 and 18) include:

I. There is repulsion of paired electrons in the p^4 configuration.
II. The electrons in a filled s orbital are more effective at shielding the electrons in the p orbitals of the same n than each other.
III. Filled orbitals and subshells are more stable than unfilled orbitals and subshells.

(A) I only
(B) II only
(C) III only
(D) I and II only
(E) I, II, and III

23. Which of the following chemical species is correctly ordered from smallest to largest radius?

(A) P < S < Cl
(B) Ne < Ar < Kr
(C) F < O < O^{2-}
(D) K < K^+ < Rb
(E) Na^+ < Mg^{2+} < Na

24. Which of the following electron configurations represents an atom in an excited state?

(A) $1s^22s^22p^5$
(B) $1s^22s^22p^53s^2$
(C) $1s^22s^22p^63s^1$
(D) $1s^22s^22p^63s^23p^5$
(E) $1s^22s^22p^63s^23p^64s^1$

Questions 25–28 refer to the ground state atoms of the following elements:

(A) Ga
(B) Tc
(C) C
(D) S
(E) N

25. This atom contains exactly one unpaired electron.

26. This atom contains exactly two unpaired electrons.

27. This atom contains exactly two electrons in the highest occupied energy sublevel.

28. This element is radioactive.

Questions 29–35 refer to the following:

(A) ↑↓

(B) ↑↓ ↑↓ ↑ __ __

(C) ↑↓ ↑↓ ↑↓ ↑↓ ↑↓ ↑

(D) ↑↓ ↑ ↑↓ ↑↓ ↑↓ ↑↓

(E) ↑↓ ↑↓ ↑ ↑ ↑

29. A highly reactive, ground state metal

30. Highest first ionization energy

31. An atom in the excited state

32. An atom that forms a trigonal planar molecule when saturated with hydrogen

33. Has exactly five valence electrons

34. The most abundant element in Earth's atmosphere

35. A chemically unreactive atom

Questions 36–42 refer to the following:

(A) __ ↑

(B) ↑↓ ↑↓

(C) ↑↓ ↑↓ ↑↓ ↑↓ ↑↓

(D) ↑↓ ↑↓ ↑↓ ↑↓ ↑↓ ↑

(E) [Ar] ↑↓ ↑↓ ↑↓ ↑ ↑ ↑

36. An atom in the excited state

37. An atom whose aqueous cation is colored

38. A chemically unreactive atom

39. An atom that forms an alkaline solution and hydrogen gas when combined with water

40. An atom with the highest second ionization energy

41. An atom that forms colored compounds

42. A highly reactive metal

43. Which of the following is the most accurate interpretation of Rutherford's experiment in which he bombarded gold foil with alpha particles?

(A) Electrons are arranged in shells of increasing energy around the nucleus of an atom.
(B) The volume of an atom is mostly empty space with the positive charges concentrated in a dense nucleus.
(C) Protons and neutrons are more massive than electrons but take up less space.
(D) Atoms are made of subatomic particles of different charges and masses.
(E) Discrete emissions spectrum lines are produced because only certain energy states of electrons are allowed.

44. All of the halogens in their element form at 25°C and 1 atm are:

(A) Gases
(B) Colorless
(C) Odorless
(D) Negatively charged
(E) Diatomic molecules

Questions 45–49 refer to the following choices:

(A) Alkali metals
(B) Noble gases
(C) Halogens
(D) Transition elements
(E) Actinides

45. The most likely to form anions

46. Their monovalent cations form clear solutions

47. Have the highest ionization energies in a given period

48. All are radioactive

49. The most difficult to oxidize in a given period

	Substance 1	**Substance 2**
Flame test	Brick red	Greenish-blue
Color in solution	Colorless	Blue

50. Two unknown, solid substances are analyzed in a lab. The results are shown above. True statements about the composition of these two substances include:

I. Substance 1 contains an alkali metal.
II. Substance 2 contains an alkali earth metal.
III. Substance 2 contains a transition metal.

(A) I only
(B) II only
(C) III only
(D) I and II only
(E) I and III only

Chemical Bonding

51. Which of the following compounds has the greatest ionic character?

(A) SiO_2
(B) ClO_2
(C) CH_4
(D) AlF_3
(E) SO_2

52. Which of the following compounds has the greatest lattice energy?

(A) KCl
(B) NaCl
(C) $CaCl_2$
(D) $MgCl_2$
(E) $FeCl_3$

53. Which of the following is the correct name for the compound with the chemical formula Mg_3N_2?

(A) Trimagnesiumdinitrogen
(B) Trimagnesiumdinitride
(C) Magnesium nitrogen
(D) Magnesium nitrate
(E) Magnesium nitride

Questions 54 and 55 refer to the following choices:

(A) N_2
(B) F_2
(C) O_3
(D) NH_3
(E) CO_2

54. Has one or more bonds with a bond order of 1.5

55. Has one or more bonds with a bond order of 3

56. If metal X forms an ionic chloride with the formula XCl_2, which of the following is most likely the formula for the stable phosphide of X?

(A) XP_2
(B) X_2P_3
(C) X_3P_2
(D) $X_2(PO_4)_3$
(E) $X_3(PO_4)_2$

57. In which of the following processes are covalent bonds broken?

(A) $C_{10}H_{8(s)} \rightarrow C_{10}H_{8(g)}$
(B) $C_{(diamond)} \rightarrow C_{(graphite)}$
(C) $NaCl_{(s)} \rightarrow NaCl_{(molten)}$
(D) $KCl_{(s)} \rightarrow KCl_{(aq)}$
(E) $NH_4NO_{3(s)} \rightarrow NH_4^+{}_{(aq)} + NO_3^-{}_{(aq)}$

58. In which of the following processes are covalent bonds broken?

(A) Solid sodium chloride melts.
(B) Bronze (an alloy of copper and tin) melts.
(C) Table sugar (sucrose) dissolves in water.
(D) Solid carbon (graphite) sublimes.
(E) Solid carbon dioxide (dry ice) sublimes.

59. Diamond is an extremely hard substance. This quality is best explained by the fact that a diamond crystal

(A) is ionic with a high lattice energy.
(B) is made completely of carbon, a very hard atom.
(C) is formed only under extremely high heat and pressure.
(D) has many delocalized electrons that contribute to greater van der Waal forces.
(E) is one giant molecule in which each atom forms strong bonds with each of its neighbors.

Questions 60–65 refer to the following answer choices:

(A) PCl_5
(B) BH_3
(C) NH_3
(D) CO_2
(E) SO_2

60. The molecule with trigonal pyramidal molecular geometry

61. The molecule with trigonal bipyramidal molecular geometry

62. The molecule with trigonal planar molecular geometry

63. The molecule with bent molecular geometry

64. The molecule with linear molecular geometry

65. The molecule with tetrahedral electron pair geometry

Questions 66–70 refer to the following answer choices:

(A) C_3H_8
(B) C_6H_6
(C) H_2O
(D) CO_2
(E) CH_2O

66. The molecule with the largest dipole moment

67. The molecule with the greatest number of π (pi) bonds

68. The molecule with the greatest number of σ (sigma) bonds

69. The molecule that contains a central atom with sp hybridization

70. The molecule with exactly one double bond

Questions 71–76 refer to the following answer choices:

(A) CO
(B) C_2H_4
(C) PH_3
(D) HF
(E) O_2

71. Has the largest dipole moment

72. Contains two π (pi) bonds

73. A necessary reactant for combustion reactions

74. Has trigonal pyramidal electron pair geometry

75. Contains the most σ (sigma) bonds

76. One of two allotropes of an element found in Earth's atmosphere

Questions 77–**80** refer to the following answer choices:

(A) CCl_4
(B) CO_2
(C) H_2O
(D) BH_3
(E) NH_3

77. This molecule has exactly 2 double bonds.

78. This molecule has the largest dipole moment.

79. This molecule has a trigonal planar molecular geometry.

80. This molecule has a trigonal pyramidal molecular geometry.

81. Types of hybridization exhibited by C atoms in ethane include which of the following?

I. sp
II. sp^2
III. sp^3

(A) I only
(B) II only
(C) III only
(D) I and III only
(E) I, II, and III

82. Types of hybridization exhibited by C atoms in hexene include which of the following?

(A) sp only
(B) sp^2 only
(C) sp^3 only
(D) sp and sp^2 only
(E) sp^2 and sp^3 only

83. Types of hybridization exhibited by C atoms in butyne include which of the following?

I. sp
II. sp^2
III. sp^3

(A) I only
(B) II only
(C) III only
(D) I and III only
(E) I, II, and III

84. There is a progressive decrease in the bond angle in the series of molecules CCl_4, PCl_3, and H_2O. According to the VSEPR model, this is best explained by:

(A) Increasing polarity of bonds
(B) Increasing electronegativity of the central atom
(C) Increasing number of unbonded electrons
(D) Decreasing size of the central atom
(E) Decreasing bond strength

85. Which of the following is a nonpolar molecule that contains polar bonds?

(A) H_2
(B) O_2
(C) CO_2
(D) CH_4
(E) CH_2F_2

86. Which of the following molecules contains *only* single bonds?

(A) C_3H_6
(B) C_6H_6
(C) C_6H_{14}
(D) CH_3CHO
(E) CH_3CH_2COOH

87. Which of the following describes the hybridization of the phosphorus atom in the compound PCl_5?

(A) sp
(B) sp^2
(C) sp^3
(D) sp^3d
(E) sp^3d^2

88. Which of the following single bonds is the *least* polar?

(A) H–N
(B) H–O
(C) F–O
(D) I–F
(E) H–F

89. Which of the following molecules has an angular (bent) geometry and is most commonly represented as a resonance hybrid of two or more Lewis-dot structures?

(A) O_3
(B) H_2O
(C) CO_2
(D) $BeCl_2$
(E) OF_2

90. Which of the following molecules has the largest dipole moment?

(A) CO
(B) HCN
(C) HCl
(D) HF
(E) NH_3

States of Matter

Questions 91–95 refer to the following descriptions of bonding in different types of solids.

(A) A lattice of closely packed cations with delocalized electrons throughout
(B) A lattice of cations and anions held together by electrostatic forces
(C) Strong, single covalent bonds connect every atom
(D) Strong, covalent bonds connect atoms within a sheet, while individual sheets are held together by weak intermolecular forces
(E) Strong, multiple covalent bonds including σ (sigma) and π (pi) bonds connect the atoms

91. Gold (Au)

92. Magnesium chloride ($MgCl_2$)

93. Carbon dioxide (CO_2)

94. Carbon ($C_{graphite}$)

95. Carbon ($C_{diamond}$)

Questions 96–98 refer to the following phase diagram of a pure substance.

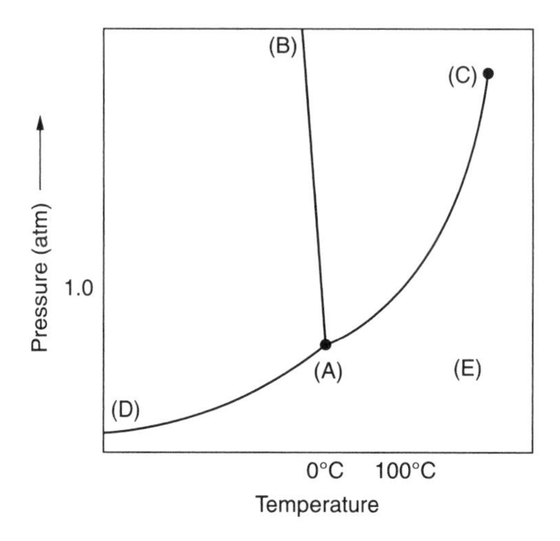

96. The point on the diagram that corresponds to the normal boiling point of the substance

97. The line on the graph that corresponds to the equilibrium between the solid and gas phases of the substance

98. All of the following are correct statements regarding the negative slope of the line indicated by the letter B *except*:

(A) As pressure increases, the temperature must decrease for the solid to form.
(B) As pressure increases, more heat must be removed from the compound in order to solidify.
(C) The freezing point of the compound is actually *lower* than the normal freezing point at pressures above 1 atm.
(D) The solid form of this compound may have a greater density than the liquid form of this compound.
(E) At low temperatures, a high pressure is required for a solid to form.

Questions 99 and 100 refer to the phase diagram for carbon dioxide.

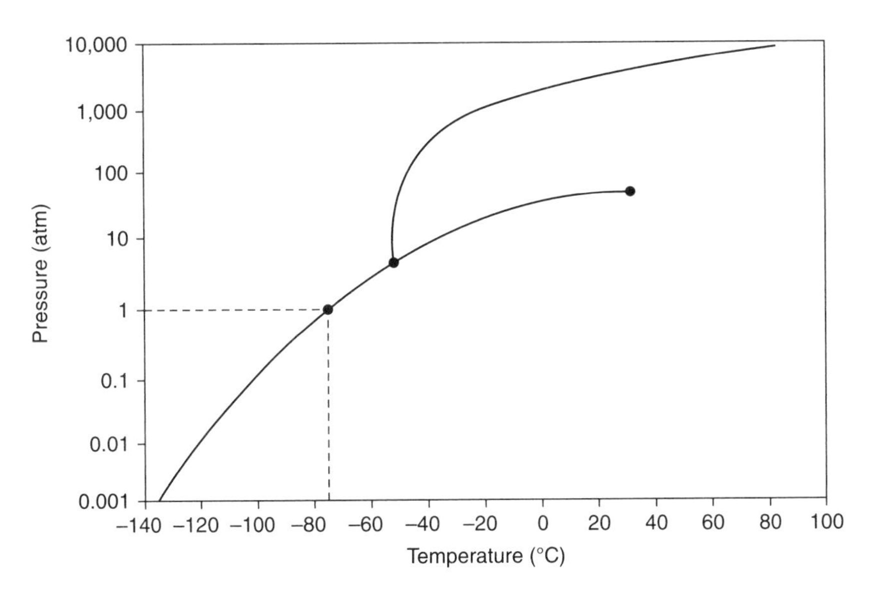

99. The temperature of a sample of pure solid is slowly raised from −100°C to 20°C at a constant pressure of 1 atm, what is the expected behavior of the substance?

(A) It melts to a liquid and then boils at −80°C.
(B) It melts to a liquid but does not boil until a temperature higher than 100°C is reached.
(C) It melts to a liquid and boils at about 30°C.
(D) It evaporates.
(E) It sublimes.

100. What is the expected behavior of the substance as the temperature is slowly raised from −100°C to −40°C at a constant pressure of 1 atm?

(A) It melts to a liquid.
(B) It sublimes to a vapor.
(C) It evaporates to a vapor.
(D) It first melts at approximately 80°C and then quickly evaporates.
(E) It first melts at approximately 80°C and then quickly sublimes.

101. Which of the following pure substances has the highest melting point?

(A) H_2S
(B) C_5H_{12}
(C) I_2
(D) SiO_2
(E) S_8

Questions 102–109 refer to the following diagram showing the temperature changes of 0.5 kg of water, starting as a solid. It is heated at a constant rate of 1 atm of pressure in an open container. Assume no mass is lost during the experiment.

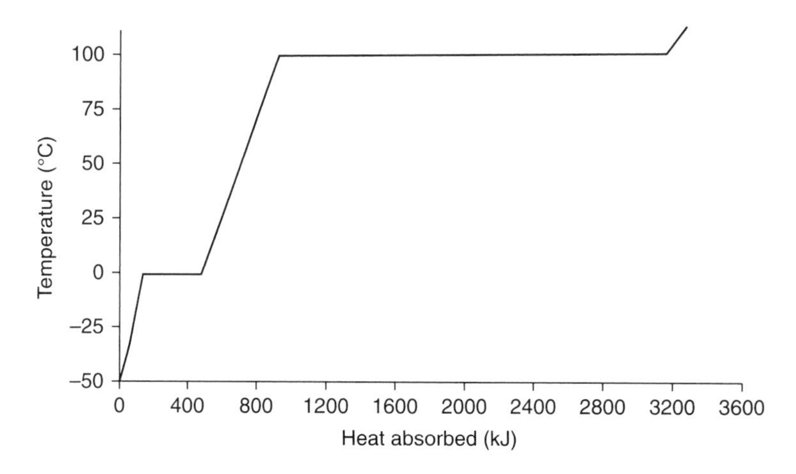

102. The sample of water requires the greatest input of energy during:

(A) The heating of ice from –50°C to 0°C
(B) The melting of ice at 0°C
(C) The heating of water from 0°C to 100°C
(D) The vaporization of water 100°C
(E) The heating of steam from 100°C to 120°C

103. Which of the following best describes what is happening at 0°C?

(A) The average kinetic energy of the particles is increasing as heat is being absorbed.
(B) The average distance between the molecules is decreasing.
(C) The number of hydrogen bonds between the molecules are increasing.
(D) The potential energy of the substance is decreasing.
(E) The substance is sublimating.

104. The heat of fusion is closest to:

(A) 75 kJ kg^{-1}
(B) 150 kJ kg^{-1}
(C) 300 kJ kg^{-1}
(D) 600 kJ kg^{-1}
(E) 750 kJ kg^{-1}

105. The heat of vaporization is closest to:

(A) 750 kJ kg^{-1}
(B) 1,500 kJ kg^{-1}
(C) 1,800 kJ kg^{-1}
(D) 2,300 kJ kg^{-1}
(E) 3,000 kJ kg^{-1}

106. The specific heat of ice is closest to:

(A) 2.0 kJ $kg^{-1}°C^{-1}$
(B) 4.2 kJ $kg^{-1}°C^{-1}$
(C) 6.0 kJ $kg^{-1}°C^{-1}$
(D) 8.4 kJ $kg^{-1}°C^{-1}$
(E) 10.0 kJ $kg^{-1}°C^{-1}$

107. How is the the disparity between the heat of fusion and the heat of vaporization best explained?

(A) It takes more hydrogen bonds for water to fuse than it does to vaporize.
(B) Water molecules are moving farther apart during fusion than during vaporization.
(C) Water molecules are moving closer together during fusion and farther apart during vaporization.
(D) Vaporization occurs at a higher kinetic energy than fusion.
(E) More hydrogen bonds are broken during vaporization.

108. The data in the heating curve graph can be used to calculate:

(A) The enthalpy of formation of water
(B) The enthalpy of hydrogen bond formation
(C) The specific heat of superheated steam
(D) The amount of time it takes for water to melt at 0°C
(E) The density of water at 50°C

109. All of the following are true regarding energy and entropy changes in the water during the experiment *except*:

(A) The energy of the water continuously increases.
(B) The kinetic energy of the water does not increase continuously.
(C) There are two points on the curve where only the potential energy and entropy of the water are increasing.
(D) The entropy of the water only increases during phase changes.
(E) The rearrangement of the water molecules during phase changes increases their potential energy.

Questions 110–112 refer to the choices in the following table.

	Compound	Vapor Pressure at 25°C (torr)
(A)	$C_2H_5OC_2H_5$	545
(B)	CS_2	336
(C)	CCl_4	115
(D)	C_2H_5OH	54
(E)	CH_3OH	122

110. The compound with the least or weakest intermolecular forces

111. The compound that can form hydrogen bonds with the strongest London dispersion forces

112. Nonpolar molecule of lowest volatility

113. The melting point of BeO is 2,507°C while the melting point of NaCl is 801°C. Explanations for this difference include which of the following?

I. Be^{2+} is more positively charged than Na^+.
II. O^{2-} is more negatively charged than Cl^-.
III. The Cl^- ion is larger than the O^{2-} ion.

(A) I only
(B) II only
(C) III only
(D) I and II only
(E) I, II, and III

114. A pure liquid heated in an open container will boil when at the temperature at which the

(A) average kinetic energy of the liquid is equal to the average kinetic energy of the gas.
(B) average kinetic energy of the liquid equals the molar entropy of the gas.
(C) entropy of the liquid equals the entropy of the gas.
(D) entropy of the vapor above the liquid equals the entropy of the atmosphere.
(E) vapor pressure of the liquid equals the atmospheric pressure above the liquid.

115. At the top of a high mountain, water boils at 90°C (instead of 100°C, the boiling point of water at sea level). Which of the following best explains this phenomenon?

(A) Water at high altitudes contains a greater concentration of dissolved gases.
(B) Water molecules at high altitudes have higher kinetic energies due to the lower pressure on them.
(C) Equilibrium because water vapor pressure equals atmospheric pressure at a lower temperature.
(D) The vapor pressure of water increases with increasing altitude.
(E) Water found at high altitudes has fewer solutes and impurities that allows boiling to occur at lower temperatures.

116. Which of the following best describes the changes that occur in the forces of attraction between CO_2 molecules as they change phase from a gas to a solid?

(A) C–O bonds are formed.
(B) Hydrogen bonds between CO_2 molecules are formed.
(C) Ionic bonds between CO_2 molecules are formed.
(D) London (dispersion) forces operate to form the solid.
(E) CO_2 molecules form a crystal around a nucleation point.

117. All of the following changes occur as H_2O freezes *except*:

(A) Ionic bonds form between the water molecules.
(B) The water takes on a crystalline structure.
(C) The density of the water decreases.
(D) The mass of the water does not change.
(E) The number of hydrogen bonds between the water molecules increases.

118. Which of the following statements accounts for the increase in boiling points of the elements going down group 18 (the noble gases)?

(A) The London (dispersion) forces increase.
(B) Atoms with a large radius are closer together.
(C) Atoms of higher mass move more slowly on average than atoms of lower mass.
(D) Dipole–dipole interactions increase.
(E) The kinetic energy of the atoms decreases with increasing mass.

119. Which of the following is expected to have the highest boiling point based on the strength of intermolecular forces?

(A) Xe
(B) Br_2
(C) Cl_2
(D) N_2
(E) O_2

120. Which of the following *must* be true of a pure, covalent solid heated slowly at its melting point until about half the compound has turned into liquid?

(A) The sum of the intermolecular forces holding the solid together decrease to zero as the solid continues to melt.
(B) Covalent bonds are broken as the solid melts.
(C) The temperature increases and the average kinetic energy of the molecules in the liquid phase increases.
(D) The volume increases as the substance becomes a liquid.
(E) The average kinetic energy of the substance remains the same.

Questions 121–127 refer to the following gases at 0°C and 1 atm.

(A) He (molar mass 4)
(B) Xe (131)
(C) O_2 (32)
(D) CO_2 (44)
(E) CO (28)

121. Has the greatest density

122. The particles (atoms or molecules) of this gas have an average speed closest to that of N_2 molecules at STP

123. Has the greatest rate of effusion

124. Has the lowest rate of effusion

125. Requires the lowest temperature and highest pressure to liquefy

126. The gas that has the greatest London dispersion forces

127. The gas that condenses at the highest temperature at 10 atm

128. Under which of the following conditions do gases behave most ideally?

(A) Low pressure and temperature
(B) High pressure and temperature
(C) High pressure, low temperature
(D) Low pressure, high temperature
(E) Any temperature if the pressure is less than 0.821 atm

129. At 298 K and 1 atm, a 0.5-mol sample of $O_{2(g)}$ and a separate 0.75-mol sample of $CO_{2(g)}$ have the same:

(A) Mass
(B) Density
(C) Average molecular speed
(D) Average molecular kinetic energy
(E) Number of atoms

130. At STP, a 0.2-mol sample of $CO_{2(g)}$ and a separate 0.4-mol sample of $N_2O_{(g)}$ have the same:

(A) Mass
(B) Volume
(C) Average molecular speed
(D) Number of atoms
(E) Chemical properties

131. The temperature at which 32.0 g of O_2 gas will occupy 22.4 L at 4.0 atm is closest to:

(A) 90 K
(B) 273 K
(C) 550 K
(D) 950 K
(E) 1,900 K

132. The pressure exerted by 2.5 mol of an ideal gas placed in a 4.00-L container at 55°C is given by which of the following expressions?

(A) $\frac{(2.5)(55.0)}{(0.0821)(4)}$ atm

(B) $\frac{(2.5)(0.0821)(323)}{4.00}$ atm

(C) $\frac{(4.00)(2.5)}{(0.0821)(55.0)}$ atm

(D) $\frac{4.00}{(2.5)(0.0821)(323)}$ atm

(E) $\frac{(2.5)(8.314)(323)}{4.00}$ atm

133. Gases $N_{2(g)}$ and $H_{2(g)}$ are added to a previously evacuated container and react at a constant temperature according to the following chemical equation:

$$N_{2(g)} + 3\ H_{2(g)} \rightarrow 2\ NH_{3(g)}$$

If the initial pressure of $N_{2(g)}$ was 1.2 atm, and that of $H_{2(g)}$ was 3.8 atm, what is the partial pressure of $NH_{3(g)}$ when the partial pressure of $N_{2(g)}$ has decreased to 0.9 atm?

(A) 0.30 atm
(B) 0.60 atm
(C) 0.9 atm
(D) 1.8 atm
(E) 3.8 atm

134. Which of the following gases behaves least ideally?

(A) Ne
(B) CH_4
(C) CO_2
(D) H_2
(E) SO_2

135. Which of the following gases will behave most ideally?

	Temperature (K)	Pressure (atm)
(A)	100	0.5
(B)	100	3.0
(C)	250	1.5
(D)	500	0.5
(E)	500	3.0

136. Equal masses of Ne and Ar are placed in a rigid, sealed container. If the total pressure in the container is 1.2 atm, what is the partial pressure of Ar?

(A) 0.20 atm
(B) 0.40 atm
(C) 0.60 atm
(D) 0.80 atm
(E) 2.40 atm

137. A flask contains 0.5 mol of $SO_{2(g)}$, 1 mol of $CO_{2(g)}$, and 1 mol of $O_{2(g)}$. If the total pressure in the flask is 750 mmHg, what is the partial pressure of $SO_{2(g)}$?

(A) 750 mmHg
(B) 375 mmHg
(C) 350 mmHg
(D) 300 mmHg
(E) 150 mmHg

138. A 2-L container will hold approximately 3 grams of which of the following gases at 0°C and 1 atm?

(A) CO_2
(B) H_2O
(C) Cl_2
(D) O_2
(E) NH_3

139. A 2-L flask contains 0.50 mole of $SO_{2(g)}$, 0.75 mole of $O_{2(g)}$, 0.75 mole of $CH_{4(g)}$, and 1.00 mole $CO_{2(g)}$. The total pressure in the flask is 800 mmHg. What is the partial pressure of $O_{2(g)}$ in the flask?

(A) 125 mmHg
(B) 188 mmHg
(C) 200 mmHg
(D) 250 mmHg
(E) 375 mmHg

140. HCl and NH_3 gases are released into opposite ends of a 1-meter (100-cm), vertical glass tube at 25° C. Their reaction quickly produces a white fog of ammonium chloride. If the two gases are released at exactly the same time, which of the following most closely approximates where the ammonium chloride fog would form?

(A) 20 cm from the side where NH_3 was released
(B) 40 cm from the side where NH_3 was released
(C) In the middle (50 cm from either side)
(D) 65 cm from the side where NH_3 was released
(E) 80 cm from the side where NH_3 was released

141. A 2-L container will hold about 7 g of which of the following gases at 0° C and 1 atm?

(A) SO_2
(B) CO_2
(C) N_2
(D) Cl_2
(E) C_4H_8

142. Which of the following gases, when collected over water, would produce the greatest yield (the highest percent collected)?

(A) CH_4
(B) HCN
(C) SO_2
(D) HCl
(E) NH_3

143. Which of the following best explains why a hot-air balloon rises?

(A) The rate of diffusion of the hot air inside the balloon is greater than the rate of diffusion of the colder air surrounding the balloon.
(B) The pressure on the walls of the balloon is greater than the atmospheric pressure.
(C) The difference in temperature and pressure between the air inside and outside the balloon creates an upward acting current.
(D) The average density of the balloon is less than that of the surrounding air.
(E) The higher pressure of the surrounding air pushes on the sides of the balloon, squeezing it up to higher altitudes.

144. A rigid metal container contains Ne gas. Which of the following is true of the gas in the tank when additional Ne is added at a constant temperature?

(A) The pressure of the gas decreases.
(B) The volume of the gas increases.
(C) The total number of gas molecules remains the same.
(D) The average speed of the gas molecules remains the same.
(E) The average distance between the gas molecules increases.

145. Equal numbers of moles of Ar(*g*), Kr(*g*), and Xe(*g*) are placed in a rigid glass vessel at room temperature. If the container has a pin hole-sized leak, which of the following will be true regarding the relative values of the partial pressures of the remaining gases after some effusion has occurred?

(A) $P_{Ar} < P_{Kr} < P_{Xe}$
(B) $P_{Xe} < P_{Kr} < P_{Ar}$
(C) $P_{Kr} < P_{Ar} < P_{Xe}$
(D) $P_{Ar} < P_{Xe} < P_{Kr}$
(E) $P_{Ar} = P_{Kr} = P_{Xe}$

146. Which of the following gases has the greatest average molecular speed at 298 K?

(A) He
(B) H_2
(C) N_2
(D) O_2
(E) Ne

Questions 147–149 refer to the following situation.

In a laboratory experiment, a student reacts Na_2CO_3 (106 g mol^{-1}) with HCl. Water displacement is used to measure the amount of CO_2 produced (the gas over water is collect in a eudiometer).

147. If the student reacts 10.6 g Na_2CO_3 in 250 ml of 2.50 M HCl, how many moles of CO_2 gas would one expect to collect?

(A) 0.10 mol CO_2
(B) 0.25 mol CO_2
(C) 0.325 mol CO_2
(D) 0.63 mol CO_2
(E) 1.625 mol CO_2

148. The volume of gas the student collects is significantly less than expected because the CO_2 gas

(A) can react with water.
(B) is denser than water vapor.
(C) has a molar mass larger than N_2 and O_2 and therefore cannot displace the air above the water in the eudiometer.
(D) has a molar mass larger than N_2 and O_2, and therefore has a lower average speed at the same temperature.
(E) is not the gas that is actually produced by the reaction.

149. The total atmospheric pressure of the laboratory (760 mmHg), as well as the temperature of the water (22°C) and the volume of gas (502 mL) in the eudiometer, are known. Which additional data, if any, is needed to calculate the number of moles of CO_2 gas collected during the experiment?

(A) The temperature of the gas collected
(B) The mass of the gas in the eudiometer
(C) The volume of $H_2O_{(l)}$ in the eudiometer
(D) The vapor pressure of water at the temperature of the water in the eudiometer
(E) No other information is needed

150. Three gases, 1.6 g He (4 g mol^{-1}), 4 g Ar (40 g mol^{-1}), and 26 g Xe (131 g mol^{-1}), are added to a previously evacuated rigid container. If the total pressure in the tank is 2.1 atm, the partial pressure of $Xe_{(g)}$ is closest to:

(A) 0.2 atm
(B) 0.3 atm
(C) 0.4 atm
(D) 0.6 atm
(E) 0.8 atm

Solutions

151. A solution is prepared by dissolving a nonvolatile solute in a pure solvent. Compared to the pure solvent, the solution

(A) has a higher normal boiling point.
(B) has a higher freezing point.
(C) has a higher vapor pressure.
(D) has less osmotic pressure.
(E) has the same vapor pressure, boiling point, and freezing point because the solute is nonvolatile.

152. A solution of NaCl is heated from 25°C to 75°C. True statements regarding this solution include which of the following?

I. The molality of the solution did not change.
II. The molarity of the solution did not change.
III. The density of the solution did not change.

(A) I only
(B) II only
(C) III only
(D) I and II only
(E) II and III only

153. Approximately what mass of $CuSO_4{\cdot}5H_2O$ (250 g mol^{-1}) is needed to prepare 125 mL of a 0.20-M copper (II) sulfate solution?

(A) 2.0 g
(B) 2.5 g
(C) 6.2 g
(D) 12.5 g
(E) 25.0 g

154. What volume of distilled water should be added to 20 mL of 5 M $HCl_{(aq)}$ to prepare a 0.8-M solution?

(A) 100 mL
(B) 105 mL
(C) 125 mL
(D) 140 mL
(E) 200 mL

155. What is the final concentration of Pb^{2+} ions when a 100 mL 0.20 M $Pb(NO_3)_2$ solution is mixed with a 100 mL 0.30 M NaCl solution?

(A) 0.005 M
(B) 0.010 M
(C) 0.015 M
(D) 0.020 M
(E) 0.025 M

156. A 0.2-M solution of K_2CO_3 is a better conductor of electricity than a 0.2-M solution of KBr. Which of the following best explains this observation?

(A) K_2CO_3 is more soluble than KBr.
(B) K_2CO_3 has more atoms than KBr.
(C) K_2CO_3 contains the carbonate ion, a polyatomic ion.
(D) KBr has a higher molar mass than K_2CO_3.
(E) KBr dissociates into fewer ions than K_2CO_3.

157. An aqueous solution that is 66 percent C_2H_4O (44 g mol^{-1}) by mass has a mole fraction of ethanol closest to:

(A) 0.29
(B) 0.44
(C) 0.50
(D) 0.66
(E) 1

158. A solution contains 144 g H_2O and 92 g of ethanol (CH_3CH_2OH, molar mass 46 g mol^{-1}). The mole fraction of ethanol is closest to:

(A) 20 percent
(B) 25 percent
(C) 40 percent
(D) 64 percent
(E) 80 percent

159. What is the molality of a solution that has 29 g NaCl dissolved in 200 g of water?

(A) 0.0025 *m*
(B) 0.025 *m*
(C) 0.15 *m*
(D) 2.5 *m*
(E) 2.9 *m*

160. Salts containing which of the following ions are insoluble in cold water?

(A) Nitrate
(B) Ammonium
(C) Sodium
(D) Phosphate
(E) Acetate

161. BaF_2 is sparingly soluble in water. The addition of dilute HF to a saturated BaF_2 solution at equilibrium is expected to

(A) raise the pH.
(B) react with BaF_2 to produce H_2 gas.
(C) increase the solubility of BaF_2.
(D) precipitate out more BaF_2.
(E) produce no change in the solution.

Questions 162–165 refer to the following solution.

Ethanol, $CH_3CH_2OH_{(l)}$, and water, $H_2O_{(l)}$, are mixed in equal volumes at 25°C and 1 atm.

162. Which of the following include endothermic processes regarding the preparation of the solution?

I. Ethanol molecules move away from other ethanol molecules as they move into solution.
II. Water molecules move away from other water molecules as they move into solution.
III. Ethanol molecules form hydrogen bonds with water molecules as they move into solution.

(A) I only
(B) II only
(C) III only
(D) I and II only
(E) I, II, and III

163. What is the mole fraction of ethanol in the solution?
(The density of ethanol and water at 25°C are 0.79 g mL^{-1} and 1.0 g mL^{-1}, respectively.)

(A) 0.24
(B) 0.33
(C) 0.40
(D) 0.50
(E) 0.72

164. Mixing different proportions of ethanol and water produce different enthalpy values. At low concentrations of water or ethanol, solvation is exothermic, but for mixing equal amounts, it is endothermic. Which of the following is a logical interpretation of this observation?

(A) The ratio of hydrogen bond breakages (between molecules of the pure liquids), and the formation of hydrogen bonds (between the two different molecules when combined in solution) varies with the ratios in which the two liquids are combined.
(B) At low concentrations of ethanol or water, fewer hydrogen bonds are formed than when mixing them in equal amounts.
(C) Mixing liquids that form the same type of intermolecular forces undergo no enthalpy changes when combined in equimolar amounts.
(D) Ethanol is capable of forming more hydrogen bonds than water.
(E) Water is capable of forming more hydrogen bonds than ethanol.

165. The intermolecular forces between ethanol and water include:

I. Hydrogen bonding
II. Dipole–dipole attraction
III. London dispersion forces

(A) I only
(B) II only
(C) III only
(D) I and III only
(E) I, II, and III

166. A 1.0-L solution contains 0.1 mol KCl, 0.1 mol $CaCl_2$, and 0.1 mol $AlCl_3$. What is the minimum number of moles of $Pb(NO_3)_2$ that must be added to precipitate all of the Cl^- ions as $PbCl_2$?

(A) 0.1 mol
(B) 0.2 mol
(C) 0.3 mol
(D) 0.4 mol
(E) 0.6 mol

167. Under which of the following sets of conditions would the most $N_{2(g)}$ be dissolved in $H_2O_{(l)}$?

	Pressure of $N_{2(g)}$ above $H_2O_{(l)}$ (atm)	Temperature of $H_2O_{(l)}$ (°C)
(A)	2.5	75
(B)	2.5	15
(C)	1	75
(D)	1	15
(E)	0.5	15

168. Sodium chloride is *least* soluble in which of the following liquids?

(A) CH_3COOH
(B) CH_3OH
(C) CCl_4
(D) H_2O
(E) HBr

169. The largest percentage of which of the following compounds can be collected by cooling a saturated solution of that compound from 90°C to 20°C?

		Solubility (g/100 g water)			
	Compound	20°C	50°C	70°C	100°C
(A)	NaCl	35	36	38	40
(B)	$KClO_3$	7	18	30	58
(C)	$K_2Cr_2O_7$	11	30	46	80
(D)	K_2SO_4	155	160	168	180
(E)	$Ce_2(SO_4)_3$	34	23	20	9

Questions 170 and 171 refer to the following data. Solutions of the five compounds in the table were mixed with equimolar solutions of one of two compounds, X or Y, also in the list. Compounds of the same identity were not combined. Assume all concentrations are 1.0 M.

	Compound 1	M	Compound 2	M	Precipitate
(A)	NaCl	1.0	X	1.0	Yes
(B)	$AgNO_3$	1.0	Y	1.0	Yes
(C)	NH_4NO_3	1.0	X	1.0	No
(D)	Li_2SO_4	1.0	X	1.0	Yes
(E)	$Mg(OH)_2$	1.0	Y	1.0	No

170. The identity of substance X

171. The identity of substance Y

172. A sample of 60 mL of 0.4 M NaOH is added to 40 mL of 0.6 M $Ba(OH)_2$. What is the hydroxide concentration [OH–] of the final solution?

(A) 0.24 M
(B) 0.40 M
(C) 0.48 M
(D) 0.50 M
(E) 0.72 M

173. A student mixes equal volumes of 1.0-M solutions of copper (II) chloride and magnesium sulfate, and no precipitate is observed. When the student mixes equal volumes of 1.0-M solutions of aluminum sulfate and copper (II) fluoride, a precipitate is observed. Which of the following is the formula of the precipitate?

(A) CuF_2
(B) $CuSO_4$
(C) AlF_3
(D) $AlCl_3$
(E) $AlSO_4$

174. Which of the following pairs of liquids forms the most ideal solution when mixed in equal volumes at 25°C?

(A) HCl and H_2O
(B) CH_3CH_2OH and H_2O
(C) CH_3CH_2OH and C_6H_{14}
(D) C_6H_{14} and C_8H_{18}
(E) C_8H_{18} and H_2O

175. Suppose a sample of a homogenous solution contains 10 percent hexane (molar mass 86 g mol^{-1}) by mass. Which of the following statements is true regarding the minimum information needed to calculate the molarity of hexane in this solution?

(A) The temperature of the solution
(B) The total mass of the solution from which the sample is taken
(C) The mass and volume of a sample of the solution
(D) The volume of the sample
(E) The mass of the sample

176. A 360-mg sample of glucose, $C_6H_{12}O_6$ (molar mass 180 g mol^{-1}), is dissolved in enough water to produce a 200-mL solution. What is the molarity of a 10-mL sample of this solution?

(A) 0.01 M
(B) 0.10 M
(C) 1.0 M
(D) 2.0 M
(E) 10.0 M

177. Which of the following aqueous solutions has the highest boiling point at 1.0 atm?

(A) 0.2 *m* NaCl
(B) 0.3 *m* $CaCl_2$
(C) 0.4 *m* K_3PO_4
(D) 0.5 *m* $NaNO_3$
(E) 0.6 *m* $C_{12}H_{22}O_{11}$

178. What is the vapor pressure of a solution in which 2.00-mol propylene glycol, a nonvolatile compound, is mixed with 8.00-mol water? Assume the solution behaves ideally and is at the temperature where the vapor pressure of water is 20.0 mmHg.

(A) 4.00 mmHg
(B) 15.00 mmHg
(C) 16.00 mmHg
(D) 18.00 mmHg
(E) 20.00 mmHg

179. A dilute hydrochloric acid solution was added to a sample of an unknown solution in a lab. A white precipitate was formed, filtered from the solution, washed with hot water, and then dissolved in a solution of NH_3. A few drops of K_2SO_4 were added to the filtrate and another white precipitate formed. What two ions were precipitated out of solution?

(A) Mg^{2+} and Pb^{2+}
(B) Mg^{2+} and Ag^+
(C) Ag^+ and Ba^{2+}
(D) Ag^+ and Pb^{2+}
(E) NH_4^+ and Pb^{2+}

180. Which of the following compounds is the *least* soluble in water?

(A) $(NH_4)_2CO_3$
(B) $BaCO_3$
(C) $Fe(NO_3)_3$
(D) $Na_3(PO_4)$
(E) LiO

Chemical Reactions

181. How many molecules are contained in 180 g of water (H_2O)?

(A) 6.02×10^{22}
(B) 1.20×10^{23}
(C) 6.02×10^{23}
(D) 1.20×10^{24}
(E) 6.02×10^{24}

182. How many oxygen atoms are in 4.4 g of CO_2?

(A) 6.02×10^{22}
(B) 1.20×10^{23}
(C) 1.20×10^{24}
(D) 6.02×10^{23}
(E) 6.02×10^{24}

183. How many atoms of hydrogen are in 1.5 g of ribose ($C_5H_{10}O_5$, 150 g mol^{-1})?

(A) 6.02×10^{22}
(B) 6.02×10^{23}
(C) 6.02×10^{24}
(D) 6.02×10^{25}
(E) 6.02×10^{26}

184. A compound contains 22.2 percent Ti, 33.3 percent C, and 44.4 percent O. What is the empirical formula for this compound?

(A) TiCO
(B) $Ti_2C_3O_4$
(C) $Ti(CO)_2$
(D) $Ti(CO)_3$
(E) $Ti(CO)_6$

185. A compound is 92 percent C and 8 percent H. What is the empirical formula for this compound?

(A) CH
(B) CH_2
(C) CH_4
(D) C_6H_6
(E) C_6H_8

186. A compound containing only carbon, hydrogen, and oxygen has a molecular mass of 150 g mol^{-1}. Which of the following may be the empirical formula of the compound?

(A) CHO
(B) C_2H_3O
(C) CH_2O
(D) CH_2O_2
(E) $C_5H_{10}O_5$

187. A compound contains 0.2 mol Pd, 0.8 mol C, 1.2 mol H, and 0.8 mol O. Which of the following is the simplest formula of this compound?

(A) $Pd_2C_8H_{12}O_8$
(B) $Pd_2(C_4H_6O_4)_2$
(C) $Pd_2(C_2H_3O_2)_3$
(D) $Pd(C_2H_3O_2)_3$
(E) $Pd(C_2H_3O_2)_2$

188. A compound contains 38 percent F and 62 percent Xe. The empirical formula of the compound is:

(A) XeF
(B) Xe_2F
(C) Xe_4F
(D) XeF_4
(E) Xe_2F_3

189. What is the empirical formula for a hydrocarbon that is 75 percent carbon by mass?

(A) CH_2
(B) CH_4
(C) CH_6
(D) C_4H
(E) C_4H_5

190. What mass of $Cu_{(s)}$ is produced when 0.050 mol Cu_2O (143 g mol^{-1}) is reduced with excess $H_{2(g)}$?

(A) 3.18 g
(B) 6.35 g
(C) 12.7 g
(D) 31.8 g
(E) 63.5 g

Questions 191–196 refer to the following answer choices:

(A) $HCl_{(aq)} + NH_{3(aq)} \rightarrow NH_4Cl_{(aq)} + H_2O_{(l)}$
(B) $Ag^+{}_{(aq)} + Cl^-{}_{(aq)} \rightarrow AgCl_{(s)}$
(C) $Mg_{(s)} + O_{2(g)} \rightarrow MgO_{2(s)}$
(D) $PtCl_{4(s)} + 2\ Cl^-{}_{(aq)} \rightarrow PtCl_6{}^{2-}{}_{(aq)}$
(E) $3\ Cl_{2(aq)} + 6\ OH^-{}_{(aq)} \rightarrow 5\ Cl^-{}_{(aq)} + ClO_3{}^-{}_{(aq)} + 3H_2O_{(l)}$

191. A reaction that produces a coordination complex

192. A reaction in which the same reactant undergoes an oxidation and a reduction

193. A neutralization reaction

194. A precipitation reaction

195. A combustion reaction

196. A reaction that produces an acidic salt

197. All of the following chemical equations represent the correct net ionic equation for the reaction that occurs when aqueous sodium hydroxide is added to a saturated solution of aluminum hydroxide *except:*

(A) $Al(OH)_3 + OH^- \rightarrow [Al(OH)_4]^-$
(B) $Al(OH)_3 + 3\ OH^- \rightarrow [Al(OH)_6]^{3-}$
(C) $Al^{3+} + 4\ OH^- \rightarrow [Al(OH)_4]^-$
(D) $Al^{3+} + 6\ OH^- \rightarrow [Al(OH)_6]^{3-}$
(E) $Al + 3\ OH^- \rightarrow Al(OH)_3$

198. Which of the following chemical equations represents the intense heating of solid hydrogen carbonate (sodium bicarbonate)?

(A) $NaHCO_3 \rightarrow Na + H_2O + CO_2$
(B) $NaHCO_3 \rightarrow Na_2CO_3 + H_2O + CO_2$
(C) $NaHCO_3 \rightarrow Na + H_2 + O_2 + CO_2$
(D) $Na(CO_3)_2 \rightarrow Na + Na_2CO_3 + H_2O + CO_2$
(E) $Na(CO_3)_2 \rightarrow Na_2CO_3 + H_2O + CO_2$

199. Which of the following chemical equations represents the reaction that occurs when pure, solid white phosphorus burns in air?

(A) $4\ P + 3\ O_2 \rightarrow 2\ P_4O_3$
(B) $4\ P + 3\ O_2 \rightarrow P_4O_6$
(C) $4\ P + 5\ O_2 \rightarrow P_4O_{10}$
(D) $P_4 + 3\ O_2 \rightarrow P_4O_6$
(E) $P_4 + O_2 \rightarrow P_4O_{10}$

200. Which of the following chemical equations represents the net ionic equation for the reaction that occurs when sodium iodide solution is added to a solution of lead (II) acetate?

(A) $2\ I^- + Pb^{2+} \rightarrow PbI_2$
(B) $Na^+ + CH_3COO^- \rightarrow NaCH_3COO$
(C) $2\ NaI + Pb(CH_3COO)_2 \rightarrow 2\ NaCH_3COO + PbI_2$
(D) $2\ NaI + Pb(CH_3COO)_2 + H_2O \rightarrow 2\ NaOH + Pb(OH)_2 + I_2$
(E) $3\ I^- + Pb^{2+} + 2\ H_2O \rightarrow PbI_2 + Pb(OH)_2 + HI$

201. All of the following pairs of substances give visible or tactile (through the production or absorption of a significant amount of heat) evidence of a chemical reaction upon mixing *except:*

(A) $HCl_{(aq)}$ and $KOH_{(aq)}$
(B) $CaCO_{3(aq)}$ and $HF_{(aq)}$
(C) $Mg_{(s)}$ and $HI_{(aq)}$
(D) $Pb(NO_3)_{2(aq)}$ and $NaCl_{(aq)}$
(E) $NH_4NO_{3(aq)}$ and $HCl_{(aq)}$

Questions 202–205 refer to the following answer choices:

(A) $MgO + CO_2 \rightarrow MgCO_3$
(B) $I^- + Cd^{2+} \rightarrow CdI_4^{2-}$
(C) $SiH_4 + O_2 \rightarrow SiO_2 + H_2O$
(D) $CaO + SO_2 \rightarrow CaSO_3$
(E) $Mg_{(s)} \rightarrow Mg^{2+} + 2\ e^-$

202. A combustion reaction

203. A reaction that produces a complex ion

204. An oxidation that is not a combustion

205. A reaction in which the product forms a gas in acidic solutions

206. When the equation for the reaction of hexane in air is correctly balanced and all coefficients are reduced to their lowest whole-number terms, the coefficient for O_2 is:

(A) 6
(B) 12
(C) 14
(D) 19
(E) 25

$$\ldots CH_3CH_2OCH_2CH_{3(g)} + \ldots O_{2(g)} \rightarrow \ldots CO_{2(g)} + \ldots H_2O_{(g)}$$

207. When the equation above is balanced and all coefficients reduced to their lowest whole number terms, the coefficient for $H_2O_{(l)}$ is:

(A) 7
(B) 6
(C) 5
(D) 3
(E) 2

$$\ldots Li_3N_{(s)} + \ldots H_2O_{(l)} \rightarrow \ldots Li^{+}{}_{(aq)} + \ldots OH^{-}{}_{(aq)} + \ldots NH_{3(g)}$$

208. When the equation above is balanced and all coefficients reduced to their lowest whole-number terms, the coefficient for $H_2O_{(l)}$ is:

(A) 1
(B) 2
(C) 3
(D) 4
(E) 6

$$\ldots Cr^{3+}{}_{(aq)} + \ldots Cl^{-}{}_{(aq)} \rightarrow \ldots Cr_{(s)} + \ldots Cl_{2(g)}$$

209. When the equation above is balanced and all coefficients reduced to their lowest whole-number terms, the coefficient for $Cl_{2(g)}$ is:

(A) 2
(B) 3
(C) 4
(D) 5
(E) 6

$$\ldots Mg_3(PO_4)_{2(s)} + \ldots H_3PO_{4(l)} \rightarrow \ldots Mg(H_2PO_4)_{2(s)}$$

210. When the equation above is balanced and all coefficients reduced to their lowest whole-number terms, the coefficient for $H_3PO_{4(l)}$ is:

(A) 1
(B) 2
(C) 3
(D) 4
(E) 5

$$\ldots CH_3CH_2OH + \ldots Cr_2O_7^{2-} + \ldots H^+ \rightarrow \ldots CH_3COOH + \ldots Cr^{3+} + \ldots H_2O$$

211. The oxidation of ethanol in an acidic solution is represented above. When the equation is balanced and all coefficients reduced to their lowest whole-number terms, the coefficient for H^+ is:

(A) 6
(B) 12
(C) 14
(D) 16
(E) 28

$$\ldots MnO_2 + \ldots OH^- + \ldots O_2 \rightarrow \ldots MnO_4^{2-} + \ldots H_2O$$

212. The reaction represented above occurs in a basic solution. When the equation is balanced and all coefficients reduced to their lowest whole-number terms, the coefficient for OH^- is:

(A) 0
(B) 1
(C) 2
(D) 4
(E) 5

Questions 213–216 refer to the following balanced chemical reaction:

$$CS_{2(l)} + 3\ O_{2(g)} \rightarrow CO_{2(g)} + 2\ SO_{2(g)}$$

213. According to the reaction above, when 0.400 mol of $CS_{2(l)}$ is reacted as completely as possible with 1.20 mol $O_{2(g)}$, the total number of moles of products is:

(A) 0.40
(B) 0.80
(C) 1.20
(D) 1.60
(E) 4.80

214. If 6.30 moles of gas are formed by the reaction indicated above, how many moles of $O_{2(g)}$ are needed to react?

(A) 1.05
(B) 2.10
(C) 4.20
(D) 6.30
(E) 9.45

215. If 33.6 L of product are formed by the above reaction at STP, how many moles of $CS_{2(l)}$ reacted?

(A) 0.50
(B) 1.00
(C) 1.50
(D) 3.00
(E) 3.36

216. An excess of Zn(s) is added to 100 mL of 0.6 M HCl at 0°C and 1 atm. What volume of gas will be produced?

(A) 67.2 mL
(B) 672 mL
(C) 1.3 L
(D) 2.24 L
(E) 6.7 L

Questions 217–219 refer to the following answer choices and the reaction represented below.

$$2\ SO_{2(g)} + O_{2(g)} \rightarrow 2\ SO_{3(g)}$$

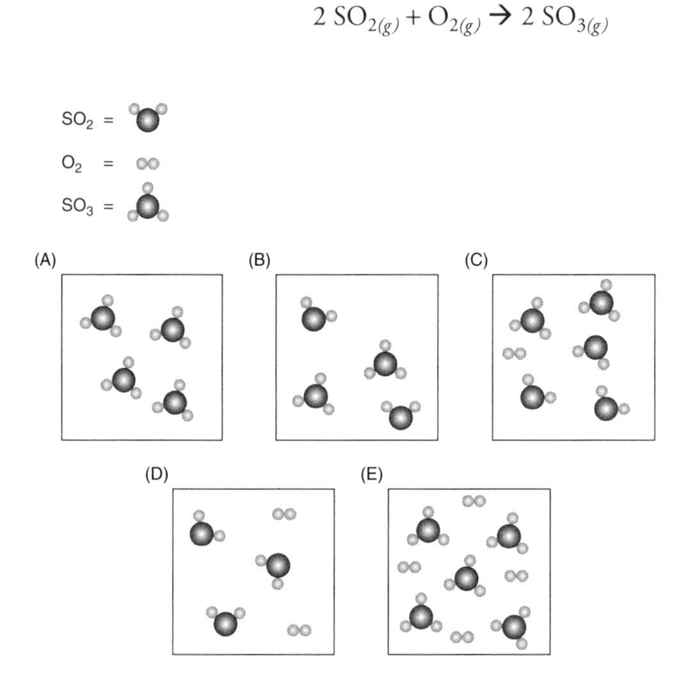

217. The reactants that would produce 3 mol SO_3 with one mol O_2 in excess

218. Products of the reaction to completion if 4 mol SO_2 combine with 1 mol O_2

219. Products if 5 mol SO_2 reacts with 6.5 mol O_2

Questions 220–224 refer to the following chemical reaction:

$$4\ NH_{3(g)} + 5O_{2(g)} \Leftrightarrow 4\ NO_{(g)} + 6\ H_2O_{(g)}$$

220. If 1.7 g NH_3 (17 g mol^{-1}) are mixed with 3.2 g O_2 (32 g mol^{-1}), what is the maximum mass of NO (30 g mol^{-1}) that can be produced?

(A) 0.40 g
(B) 2.4 g
(C) 3.2 g
(D) 4.0 g
(E) 4.8 g

221. Suppose NH_3 and O_2 were mixed in a 4:5 molar ratio. When the reaction was completed, 228 grams of the reaction gases were recovered. Assuming the reaction mixture reacted completely and the gases collected contain 100 percent products, what masses of NO and H_2O are expected to be present?

	NO (g)	H_2O (g)
(A)	57.00	38.00
(B)	57.00	45.60
(C)	68.00	160.0
(D)	91.2	136.8
(E)	120.0	108.0

222. If 100 mL of distilled water is added to 400 mL of 0.375 M NaCl, which of the following is closest to the Na^+ concentration in the final solution?

(A) 0.15 M
(B) 0.19 M
(C) 0.30 M
(D) 0.33 M
(E) 0.69 M

223. How many moles of oxygen gas would be required to produce 1.0 L of NO at STP?

(A) $\frac{5}{4}$ mol

(B) $\frac{4}{5}$ mol

(C) $\frac{1}{22.4}$ mol

(D) $\frac{4}{(22.4)(5)}$ mol

(E) $\frac{5}{(22.4)(4)}$ mol

224. If 2.5 mol NH_3 reacted with 2.5 mol O_2 as completely as possible, how many moles of reactant would remain unreacted (in excess)?

(A) 0.50 mol O_2
(B) 0.50 mol NH_3
(C) 0.63 NH_3
(D) 2.0 mol O_2
(E) 2.0 mol NH_3

Questions 225–227 refer to the following reaction of potassium superoxide (KO_2). KO_2 is used for its ability to absorb carbon dioxide gas and release oxygen.

$$4\ KO_{2(s)} + 2\ CO_{2(g)} \rightarrow 2\ K_2CO_{3(s)} + 3\ O_{2(g)}$$

225. According to the equation above, how many moles of $O_{2(g)}$ can be released if 6.00 mol KO_2 and 9.00 mol of $CO_{2(g)}$ are available?

(A) 4.50
(B) 6.00
(C) 9.00
(D) 12.0
(E) 13.5

226. What volume of CO_2 is required to react with excess KO_2 to produce 6 L of $O_{2(g)}$ at STP?

(A) 4.00 L
(B) 11.2 L
(C) 14.9 L
(D) 22.4 L
(E) 89.6 L

227. When 2.9 g $CO_{2(g)}$ (molar mass 44 g mol^{-1}) reacts with excess KO_2 according to the equation above, the volume of $O_{2(g)}$ produced at STP is closest to:

(A) 2.2 L
(B) 4.4 L
(C) 11.2 L
(D) 22.4 L
(E) 33.6 L

228. Which of the following is the correct net ionic equation for the reaction of sodium hydroxide and nitric acid?

(A) $H^+ + OH^- \rightarrow H_2O$
(B) $Na^+ + NO_3^- \rightarrow NaNO_3$
(C) $NaOH + HNO_3 \rightarrow NaNO_3 + H_2O$
(D) $Na^+ + OH^- + H^+ + NO_3^- \rightarrow NaNO_3 + H_2O$
(E) $Na^+ + OH^- + 2\ H^+ \rightarrow NaOH + H_2$

229. Which of the following is an addition reaction?

(A) $CH_4 + Cl_2 \rightarrow CH_3Cl + HCl$
(B) $CH_2{=}CH_2 + Cl_2 \rightarrow CH_2ClCH_2Cl$
(C) $C_4H_{10} + Cl_2 \rightarrow C_4H_{10}Cl + HCl$
(D) $CH_3(CH_2)_6CH{=}CHCOOH + NaOH \rightarrow$
$CH_3(CH_2)_6CH{=}CHCOONa + H_2O$
(E) $6\ CO_2 + 6\ H_2O \rightarrow C_6H_{12}O_6 + 6\ O_2$

230. Which of the following is a product of the reaction between carbon dioxide and water, $CO_{2(g)} + H_2O_{(l)}$?

I. H^+
II. CO_3^{2-}
III. HCO_3^-

(A) I only
(B) I and II only
(C) I and III only
(D) II and III only
(E) I, II, and III

Thermodynamics

Questions 231–237 refer to the following answer choices:

(A) Activation energy
(B) Lattice energy
(C) Free energy
(D) Kinetic energy
(E) Potential energy

231. The energy needed to convert an ionic solid into well-separated gaseous ions

232. The energy liberated from a physical or chemical process that is available to do work

233. The energy needed for the formation of the transition state of a chemical reaction

234. The energy needed to overcome the activation barrier of a chemical reaction

235. This quantity is determined by measuring the rate of a particular reaction at two or more different temperatures

236. This quantity must change in any substance undergoing a phase change

237. The formula for this quantity is used to derive the effusion ratio of O_2 and CO_2 at the given temperature

Questions 238–241 refer to the following answer choices:

(A) Enthalpy of formation
(B) Entropy
(C) Energy of crystallization
(D) Activation energy
(E) Gibbs free energy

238. A quantity that is zero for a perfect, pure crystalline solid at 0 K

239. A quantity that is zero when a reaction is at equilibrium at constant temperature and pressure

240. A quantity that is zero for a pure element in its standard state

241. A quantity that describes a feature of a chemical reaction and is always positive for reactions in which an increase in temperature results in an increased reaction rate

Questions 242–247 refer to the following answer choices:

(A) $\Delta H > 0, \Delta S > 0$
(B) $\Delta H < 0, \Delta S < 0$
(C) $\Delta H > 0, \Delta S < 0$
(D) $\Delta H < 0, \Delta S > 0$
(E) $\Delta H = 0, \Delta S > 0$

242. Must be true for a reaction that is spontaneous at all temperatures

243. True for melting of water at 25°C and 1 atm

244. True for the deposition of $CO_{2(g)}$ into $CO_{2(s)}$ at –100°C and 1 atm

245. True for the combustion of wood into $CO_{2(g)}$ and $H_2O_{(g)}$ in a campfire

246. Must be true for a reaction that is never spontaneous at any temperature

247. True for the mixing of two ideal gases (no intermolecular forces of attraction or repulsion occur between any of the gas particles) at a constant temperature and pressure. Assume no reaction occurs.

Compound	ΔH°_f (kJ mol^{-1})
$CO_{2(g)}$	–393.5
$CaO_{(s)}$	–635.5
$CaCO_{3(s)}$	–1,207.1

$$CaCO_{3(s)} \rightarrow CaO_{(s)} + CO_{2(g)}$$

248. The decomposition of $CaCO_{3(s)}$ is shown in the equation above. Using the data in the table above the reaction, which of the following values is closest to the ΔH_{rxn} of the decomposition of $CaCO_{3(s)}$?

(A) –2,240 kJ mol^{-1}
(B) –180 kJ mol^{-1}
(C) 180 kJ mol^{-1}
(D) 1,207 kJ mol^{-1}
(E) 2,240 kJ mol^{-1}

$$C_{(diamond)} + O_{2(g)} \rightarrow CO_{2(g)} \Delta H = -395.4 \text{ kJ}$$

$$C_{(graphite)} + O_{2(g)} \rightarrow CO_{2(g)} \Delta H = -393.5 \text{ kJ}$$

$$C_{(graphite)} \rightarrow C_{(diamond)} \Delta H_{rxn} = ?$$

249. The reactions for the combustion of diamond and graphite are shown above. Which of the following values is closest to the ΔH_{rxn} for the conversion of $C_{(graphite)}$ to $C_{(diamond)}$?

(A) –789 kJ
(B) –1.9 kJ
(C) 1.9 kJ
(D) 394.45 kJ
(E) 798 kJ

$$CaCl_{2(s)} \rightarrow Ca^{2+}_{(aq)} + 2\ Cl^{-}_{(aq)}$$

250. The entropy change for the dissolution of calcium chloride in water shown above might be expected to be positive, but the actual ΔS is negative. Which of the following is the most plausible explanation for the net loss of entropy during this process?

(A) $CaCl_{2(s)}$ is an amorphous solid.
(B) The particles in the solution are more ordered than the particles in the solid.
(C) The ions in solution can move more freely than the particles in the solid.
(D) The decreased entropy of the water molecules in the solution is greater than the increased entropy of the ions.
(E) In solution, the distance between ions is much greater than the distance between the ions in the solid.

251. Which of the following best describes the role of a spark in a butane lighter?

(A) The spark decreases the activation energy of the combustion reaction.
(B) The spark increases the concentration of butane in the reaction chamber.
(C) The spark provides the heat of vaporization for the liquid fuel.
(D) The spark supplies some of the energy to form the activated complex for the combustion reaction.
(E) The spark provides an alternative stoichiometry for the reaction, decreasing the amount of oxygen required for complete combustion.

252. Which of the following processes demonstrates a decrease in entropy ($\Delta S < 0$)?

(A) $Br_{(s)} \rightarrow Br_{(l)}$
(B) $I_{2(s)} \rightarrow I_{2(g)}$
(C) Combining equal volumes of $C_2H_6O_{2\,(l)}$ and $H_2O_{(l)}$
(D) The precipitation of PbI_2 from solution
(E) The thermal expansion of a helium balloon

253. Which of the following reactions involves the largest decrease in entropy?

(A) $MgCO_{3(s)} \rightarrow MgO_{(s)} + CO_{2(g)}$
(B) $3\ H_{2(g)} + N_{2(g)} \rightarrow 2\ NH_{3(g)}$
(C) $4\ La_{(s)} + 3\ O_{2(g)} \rightarrow 2\ La_2O_{3(s)}$
(D) $2\ NaI_{(aq)} + Pb(CH_3COO)_{2(aq)} \rightarrow 2\ NaCH_3COO_{(aq)} + PbI_{2(s)}$
(E) $C_3H_{8(g)} + 5\ O_{2(g)} \rightarrow 3\ CO_{2(g)} + 4\ H_2O_{(g)}$

$$2\ CH_6N_{2(l)} + 5\ O_{2(g)} \rightarrow 2\ N_{2(g)} + 2\ CO_{2(g)} + 6\ H_2O_{(g)}$$

254. The combustion of methylhydrazine, a common rocket fuel, is represented above. The ΔH of this reaction is $-1{,}303$ kJ $mol^{-1}CH_6N_{2(l)}$. What would be the ΔH per mol $CH_6N_{2(l)}$ if the reaction produced $H_2O_{(l)}$ instead of $H_2O_{(g)}$? (The ΔH for the condensation of $H_2O_{(g)}$ to $H_2O_{(l)}$ is -44 kJ mol^{-1}.)

(A) $-1{,}171$ kJ
(B) $-1{,}259$ kJ
(C) $-1{,}347$ kJ
(D) $-1{,}567$ kJ
(E) $-1{,}435$ kJ

255. A student adds solid ammonium chloride to a beaker containing water at 25°C. As it dissolves, the beaker feels colder. Which of the following are true regarding the ΔH and ΔS of the dissolution process of $NH_4Cl_{(s)}$?

	ΔH	ΔS
(A)	+	+
(B)	+	–
(C)	0	0
(D)	–	+
(E)	–	–

256. Suppose a reaction is spontaneous at temperatures only below 300 K. If the ΔH° for this reaction is –18.0 kJ mol^{-1}, the value of ΔS° for this reaction is closest to which of the following? Assume ΔS° and ΔH° do not change significantly with temperature.

(A) –60 J $mol^{-1}K^{-1}$
(B) –18 J $mol^{-1}K^{-1}$
(C) –0.0010 J $mol^{-1}K^{-1}$
(D) 18 J $mol^{-1}K^{-1}$
(E) 18,000 J $mol^{-1}K^{-1}$

$$Ag_{(s)} \Leftrightarrow Ag_{(l)}$$

257. The normal melting point of $Ag_{(s)}$ is 962°C. Which of the following is true for the process represented above at 962°C?

(A) $\Delta H = 0$
(B) $\Delta S = 0$
(C) $T\Delta S = 0$
(D) $\Delta H = T\Delta S$
(E) $\Delta H > T\Delta S$

258. The standard Gibbs free energy change, $\Delta G°_{298}$, for the conversion of $C_{diamond}$ into $C_{graphite}$ has an approximate value of –3 kJ mol^{-1}. However, graphite does *not* form from diamond under standard conditions (298 K and 1 atm). Which of the following best explains this observation?

(A) Diamond is more ordered than graphite (lower entropy).
(B) The ΔH for the conversion of diamond to graphite is highly endothermic.
(C) The activation energy for the conversion of diamond to graphite is very large.
(D) The C–C bonds in diamond are much stronger than the C–C bonds in graphite.
(E) Diamond is significantly denser than graphite.

259. Which of the following is true regarding the adiabatic and reversible compression of an ideal gas?

(A) The temperature of the gas remains constant.
(B) The volume of the gas remains constant.
(C) The pressure of the gas remains constant.
(D) No work can be done by the gas.
(E) The net entropy change of the gas is zero.

Bond	Energy (kJ mol^{-1})
H–H	432
O=O	494
H–O	459

260. Given the bond energies in the table above, which of the following statements best describes the formation of 1 mole of $H_2O_{(l)}$ from $H_{2(g)}$ and $O_{2(g)}$?

(A) The process is endothermic with an enthalpy change of approximately 480 kJ.
(B) The process is endothermic with an enthalpy change of approximately 240 kJ.
(C) The process is exothermic with an enthalpy change of approximately 480 kJ.
(D) The process is exothermic with an enthalpy change of approximately 240 kJ.
(E) The process is exothermic with an enthalpy change of approximately 1,800 kJ.

$$C_2H_5OH_{(l)} + 3\ O_{2(g)} \rightarrow 2\ CO_{2(g)} + 3\ H_2O_{(l)} \Delta H_{rxn} = -1{,}367\text{ kJ}$$

Compound	$\Delta H°_f$
$C_2H_5OH_{(l)}$	–278 kJ
$H_2O_{(l)}$	–286 kJ
$CO_{2(g)}$	?

261. The enthalpy of combustion of ethanol ($C_2H_5OH_{(l)}$) is shown above. Using this information and the data in the table above, the standard heat of formation of $CO_{2(g)}$ is closest to:

(A) –1,080 kJ mol^{-1}
(B) –540 kJ mol^{-1}
(C) –510 kJ mol^{-1}
(D) –390 kJ mol^{-1}
(E) –250 kJ mol^{-1}

$$\mathbf{2\ NO_{2(g)} \rightarrow N_2O_{4(g)}} \qquad \Delta H^\circ_f \text{ of } NO_2 = 34 \text{ kJ mol}^{-1}$$

$$\Delta H^\circ_f \text{ of } N_2O_4 = 9.7 \text{ kJ mol}^{-1}$$

262. What is the standard enthalpy change, ΔH°_{rxn}, of the reaction represented above?

(A) –24.3 kJ
(B) –58.3 kJ
(C) 24.3 kJ
(D) 58.3 kJ
(E) 77.7 kJ

$$CaCO_{3(s)} + 2\ HCl_{(aq)} \rightarrow CO_{2(g)} + CaCl_{2(aq)} + H_2O_{(l)}$$

263. When $CO_{3(s)}$ is added to a beaker of dilute $HCl_{(aq)}$ at 298 K, the beaker gets cold and a gas is produced. Which of the following indicates the correct signs for ΔG, ΔH, and ΔS for the reaction represented above?

	ΔG	ΔH	ΔS
(A)	–	–	–
(B)	–	–	+
(C)	+	–	–
(D)	–	+	+
(E)	+	+	+

$$A + B \rightarrow C \qquad E_a = 5.3 \text{ kJ mol}^{-1}$$

$$C \rightarrow A + B \qquad E_a = 5.3 \text{ kJ mol}^{-1}$$

264. Considering the data in the table above, which of the following must be true of the reaction?

(A) A catalyst is present.
(B) The reaction order is zero.
(C) The reaction is at equilibrium.
(D) The enthalpy change of the reaction is zero.
(E) This reaction is occurring at a temperature above 298 K.

Questions 265–269 refer to the data in the following table.

Reaction	Equation	ΔH°_{298}	ΔS°_{298}	ΔG°_{298}
X	$C_{(s)} + H_2O_{(g)} \Leftrightarrow CO_{(g)} + H_{2(g)}$	+131 kJ mol^{-1}	+134 J mol^{-1} K	+91 kJ mol^{-1}
Y	$CO_{2(g)} + H_{2(g)} \Leftrightarrow H_2O_{(g)} + CO_{(g)}$	+41 kJ mol^{-1}	+42 J mol^{-1} K	+29 kJ mol^{-1}
Z	$2\ CO_{(g)} \Leftrightarrow C_{(s)} + CO_{2(g)}$	??	??	??

265. What is the value of ΔH°_{298} for reaction Z?

(A) 90 kJ mol^{-1}
(B) –90 kJ mol^{-1}
(C) 172 kJ mol^{-1}
(D) –172 kJ mol^{-1}
(E) 213 kJ mol^{-1}

266. Reactions for which the value of K_p will *increase* under greater pressure include which of the following?

(A) X only
(B) Z only
(C) X and Z
(D) The values of K_p decrease with increased pressure
(E) None of the K_p values will increase

267. Reactions for which the value K_p will *increase* if the temperature is raised above 298 K include which of the following?

(A) X only
(B) Y only
(C) Z only
(D) X and Y only
(E) Y and Z only

268. Which of the following most accurately describes the ΔS of reaction Z?

(A) Positive, because there are more products than reactants.
(B) Positive, because there are more states of matter in the products.
(C) Positive, because there is only one species of reactant but there are two species of product.
(D) Negative, because two moles of gas are converted to a solid and one mole of gas.
(E) Negative, because a pure element was formed.

269. Which of the following statements most accurately describes the rates of reactions X and Y?

(A) X will occur more rapidly than Y because the ΔH is more positive.
(B) Y will occur more rapidly than X because the ΔH is less positive.
(C) X will occur more rapidly than Y because the ΔS and ΔG are more positive.
(D) Y will occur more rapidly than X because the ΔS and ΔG are less positive.
(E) Thermodynamic data for overall reactions do not indicate anything about the rate of a chemical reaction.

Kinetics

Questions 270–273 refer to the following choices:

(A) Rate = k [M]
(B) Rate = k [M][N]
(C) Rate = k $[M][N]^2$
(D) Rate = k $[M]^2[N]^2$
(E) Rate = k $[N]^2$

270. Doubling the concentration of M has no effect on the reaction rate.

271. Doubling the concentration of M and N increases reaction rate by 2.

272. Doubling the concentration of M only quadruples the reaction rate.

273. Doubling the concentration of M and N increases the reaction rate by eight fold. Halving the concentration of N decreases the reaction rate four fold.

274. The production of iron (II) sulfide occurs at a significantly higher rate when iron filings are used instead of blocks (volume = 0.1 mL). Which of the following best explains this observation?

(A) The iron filings are partially oxidized due to their greater exposure to oxygen.
(B) The iron in the block is $Fe_{(s)}$ and the iron in the filings is Fe^{2+}.
(C) The iron block is too concentrated to chemically react.
(D) The iron filings have a much greater area in contact with sulfur.
(E) The reactant order of iron in the rate law is 1 for $Fe_{(s)}$ and 2 for Fe^{2+}.

275. A chemistry student sitting around a campfire observes that the large pieces of wood burn slowly, but a mixture of small scraps of wood and sawdust added to the flame combusts explosively. The correct explanation for the difference in the combustion between these two forms of wood is that, compared with the wood scraps and sawdust, the large pieces of wood

(A) have a greater surface area-to-volume ratio.
(B) have a smaller surface area per kilogram.
(C) have a higher percent carbon.
(D) contain compounds with a lower heat of combustion.
(E) contain more carbon dioxide and water.

276. All of the following result in an increased rate of reaction in an aqueous solution *except*:

(A) Increasing the temperature of an endothermic reaction.
(B) Increasing the temperature of an exothermic reaction.
(C) Increasing the surface area of a solid reactant.
(D) Increasing the pressure on the solution.
(E) Mixing or stirring the solution.

277. Factors that affect the rate at which a chemical reaction proceeds include which of the following?

I. The orientation of the reactants at time of collision
II. The kinetic energy of the collisions between reactants
III. The frequency of collisions between reactants

(A) I only
(B) II only
(C) III only
(D) I and II only
(E) I, II, and III

278. Which of the following correctly explain(s) the effect of increased temperature on the rate of a chemical reaction?

I. Increases the reaction rate of endothermic reactions
II. Increases the reaction rate of exothermic reactions
III. Decreases the reaction rate of reactions with a $-\Delta H$

(A) I only
(B) II only
(C) I and II only
(D) II and III only
(E) I and III only

279. All of the following statements regarding the kinetics of radioactive decay are true *except*:

(A) The length of time of a half-life is specific to a particular element.
(B) All radioactive decay displays first-order kinetics.
(C) In a sample of a pure, radioactive isotope, one-half the number of radioactive atoms and one-half the mass of the radioactive substance remains after one half-life.
(D) The half-life of an atom does not change when the atom is incorporated into a compound.
(E) The half-life of a particular substance does not change with time or temperature.

Questions 280–285 refer to the reaction of nitrogen monoxide (nitric oxide) and oxygen and the following data.

$$2\ NO + O_2 \rightarrow 2\ NO_2$$

	[NO] (mol L^{-1})	$[O_2]$ (mol L^{-1})	RATE (M s^{-1})
Trial 1	2.4×10^{-2}	3.5×10^{-2}	1.43×10^{-1}
Trial 2	1.5×10^{-2}	3.5×10^{-2}	5.6×10^{-2}
Trial 3	2.4×10^{-2}	4.5×10^{-2}	1.84×10^{-1}

280. The rate law for this reaction is:

(A) Rate = k $[NO][O_2]$
(B) Rate = k $[NO][O_2]^2$
(C) Rate = k $[NO]^2[O_2]$
(D) Rate = k $[NO]^2[O_2]^2$
(E) Rate = k $[NO][O_2]^{1.25}$

281. The numerical value for the rate constant (k) is closest to:

(A) 5.9×10^{-2}
(B) 170
(C) 3×10^3
(D) 7×10^3
(E) 1.2×10^5

282. The unit of the rate constant (k) is:

(A) sec^{-1}
(B) L mol^{-1} sec^{-1}
(C) L^2mol^{-2} sec^{-1}
(D) L^3mol^{-3} sec^{-1}
(E) L^4mol^{-4} sec^{-1}

283. Increasing the initial concentration of NO fivefold would increase the reaction rate by:

(A) 5 X
(B) 10 X
(C) 25 X
(D) 3,125 X
(E) No substantial margin

284. What would be the reaction rate if the initial concentration of NO was 2×10^{-2} M and the initial concentration of O_2 was 4×10^{-2} M?

(A) 3.2×10^{-5}
(B) 8×10^{-4}
(C) 2.3×10^{-2}
(D) 1.1×10^{-1}
(E) 5.7

285. Which of the following is a correct statement about reaction order?

(A) Reaction order must be a whole number.
(B) Reaction order can be determined mathematically using only coefficients of the balanced reaction equation.
(C) Reaction order can change with increasing temperature.
(D) A second-order reaction must involve at least two reactants.
(E) Reaction order can only be determined experimentally.

286. Properties of a catalyst include all of the following *except:*

(A) A catalyst that works for one chemical reaction may not work for a different reaction.
(B) A catalyst is not consumed by the reaction it catalyzes.
(C) Catalysts can be solids or gases.
(D) A catalyst will only speed up a chemical reaction in *either* the forward *or* reverse direction.
(E) Catalysts speed up chemical reactions by providing an alternate pathway for reaction in which the activated complex is of lower energy.

287. Which of the following can be used to calculate or measure the rate of a chemical reaction?

I. The appearance of product over time
II. The disappearance of one or more substrates over time
III. The rate law
IV. The K_{eq} (equilibrium constant) and Q (reaction quotient) of the reaction

(A) I and II only
(B) I and III only
(C) I, II, and III only
(D) I, II, and IV only
(E) I, II, III, and IV

288. The reaction of nitric oxide with hydrogen gas at 25°C and 1 atm is represented below. The rate law for this reaction is: rate = k $[H_2][NO]^2$.

$$2\ NO_{(g)} + 2\ H_{2(g)} \rightarrow N_{2(g)} + 2\ H_2O_{(g)}$$

According to the rate law, which of the following is the best prediction of the rate of this reaction?

(A) The rate of disappearance of NO is always twice as great as the disappearance of H_2.
(B) The rate of disappearance of NO is always four times as great as the disappearance of H_2.
(C) The rate of disappearance of NO is twice as great as the disappearance of H_2 if the concentration of NO is initially twice that of H_2.
(D) The rate of disappearance of NO is four times as fast as that of H_2, but only if the initial concentration of NO is initially twice that of H_2.
(E) The relative disappearances of NO and H_2 cannot be deduced without the value of the rate constant (k).

$$NO_{2(g)} + CO_{(g)} \rightarrow NO_{2(g)} + CO_{2(g)}$$

289. The reaction between nitrogen dioxide and carbon monoxide is represented above. The proposed reaction mechanism is as follows:

$$\text{Step 1: } NO_2 + NO_2 \rightarrow NO_3 + NO \text{ (slow)}$$

$$\text{Step 2: } NO_3 + CO \rightarrow NO_2 + CO_2 \text{ (fast)}$$

Which of the following reaction mechanisms is consistent with the proposed mechanism?

(A) Rate = k $[NO_2]$
(B) Rate = k $[NO_2]^2$
(C) Rate = k $[NO][CO]$
(D) Rate = k $[NO]^2[CO]$
(E) Rate = k $[CO]$

$$2\ NO_{(g)} + Br_{2(g)} \rightarrow 2\ NOBr_{(g)}$$

290. The reaction between nitrogen monoxide (commonly called nitric oxide) and bromine is represented above. The proposed reaction mechanism is as follows:

$$NO + Br_2 \Leftrightarrow NOBr_2 \text{ (fast)}$$

$$NOBr_2 + NO \rightarrow 2\ NOBr \text{ (slow)}$$

Which of the following reaction mechanisms is consistent with the proposed mechanism?

(A) Rate = k $[NO]^2$
(B) Rate = k $[NO][Br_2]$
(C) Rate = k $[NO][Br_2]^2$
(D) Rate = k $[NO]^2[Br_2]$
(E) Rate = k $[NO]^2[Br_2]^2$

291. The rate constant for a certain chemical reaction at 25°C is 9.0×10^5 $L^2mol^{-2}sec^{-1}$. Which of the following must be true regarding this reaction?

(A) This reaction is slower than a reaction that has a rate constant of $9.0\ L^2mol^{-2}sec^{-1}$.
(B) This reaction is exothermic.
(C) The rate of this reaction will decrease with increasing temperature.
(D) The reaction order is 3.
(E) Doubling the concentration of reactants will increase the reaction rate by a factor of 8.1×10^{10}.

292. The conversion of ozone, $2\ O_{3\,(g)} \rightarrow 3\ O_{2\,(g)}$, obeys the rate law, rate = k $[O_3]^2[O_2]^{-1}$. Which of the following statements is true regarding the rate of the breakdown of ozone (O_3) into molecular oxygen (O_2)?

(A) The catalyst that converts ozone to oxygen is inhibited by oxygen.
(B) The rate at which ozone is converted to oxygen increases as the concentration of oxygen decreases.
(C) There is an inverse square relationship between ozone and oxygen concentrations.
(D) The conversion of oxygen to ozone is faster than the conversion of ozone to oxygen.
(E) The negative reactant order of oxygen indicates that the reaction is at equilibrium and the forward reaction is more favorable.

293. All of the following statements regarding the activated complex are true *except:*

(A) The energy of the activated complex determines the activation energy of the reaction.
(B) The activated complex represents the highest energy state along the transition path of a chemical reaction.
(C) The activated complex of a chemical reaction is specific to that reaction.
(D) The configuration of atoms in the activated complex of an uncatalyzed reaction is the same as that of a catalyzed reaction, except for the presence of the catalyst.
(E) A reaction in which the energy of the activated complex is very large indicates that the reactants are very thermodynamically stable (as opposed to chemically stable).

Questions 294 and 295 refer to the following graph of a chemical reaction over time.

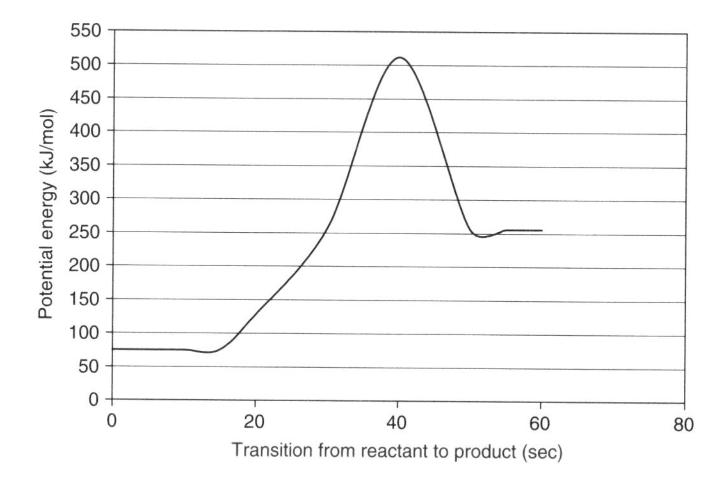

294. Which of the following is true concerning the reaction at 25°C?

I. The reaction is endothermic.

II. The activation energy (E_a) is approximately 510 kJ mol^{-1}.

III. The magnitude of the difference between energy of the reactants and products is approximately 175 kJ mol^{-1}.

(A) I only
(B) II only
(C) III only
(D) II and III only
(E) I and III only

295. Which of the following is true concerning the effect of adding a catalyst?

(A) The activated complex would form in less than 40 seconds.
(B) It lowers the energy of the products.
(C) It increases the energy of the reactants.
(D) More reactions would occur per second.
(E) The equilibrium would shift to favor the products.

Time (minutes)	0	5	10	15	20	25	30
Reactant Remaining *(g)*	100	75	56	42	32	24	18

296. A reaction was observed for 30 minutes. Every 5 minutes, the percent of reactant remaining was measured. According to the data in the table above, which of the following most accurately describes the reaction order and half-life of this reaction?

	Reaction Order	**Half-Life (minutes)**
(A)	Zero	10
(B)	First	12
(C)	First	8
(D)	Second	12
(E)	Second	8

297. The rate laws for the reaction between O_2 and NO is $k = [O_2][NO]^2$. If the reaction rate is first measured with O_2 and NO of 2.5×10^{-4} M each, by what factor will the rate increase if the concentration of O_2 and NO are both increased to 5.0×10^{-4} each?

(A) 2
(B) 3
(C) 4
(D) 6
(E) 8

Questions 298 and 299 refer to the data below.

Trial	[X]	[Y]	Formation of Product Z
1	0.5	0.1	R
2	0.25	0.2	?

298. Consider the data shown above. If the data were obtained from a reaction whose rate law is $k = [X][Y]^2$, what would be the expected rate of reaction for Trial 2?

(A) R
(B) 2R
(C) 4R
(D) $\frac{R}{2}$
(E) $\frac{R}{4}$

299. Suppose the data were obtained for a reaction whose rate law is $k = [X]^2[Y]$. What would be the expected rate of reaction for Trial 2?

(A) R
(B) 2R
(C) 4R
(D) $\frac{R}{2}$
(E) $\frac{R}{4}$

300. The units of the rate constant (k) for a reaction that occurs between two second-order reactants is:

(A) sec^{-1}
(B) $L\ mol^{-1}\ sec^{-1}$
(C) $L^2\ mol^{-2}\ sec^{-1}$
(D) $L^3\ mol^{-3}\ sec^{-1}$
(E) $L^4\ mol^{-4}\ sec^{-1}$

Equilibrium

$$2\ Mg_{(s)} + O_{2(g)} \Leftrightarrow 2\ MgO_{(s)}$$

301. Which of the following stresses when applied to the reaction above will result in an increased amount of $MgO_{(s)}$?

I. Remove $Mg_{(s)}$
II. Increase the pressure (decrease the volume)
III. Add $O_{2(g)}$

(A) I only
(B) II only
(C) III only
(D) II and III only
(E) I, II, and III

$$N_{2(g)} + O_{2(g)} + Cl_{2(g)} + 104\ kJ \Leftrightarrow 2\ NOCl_{(g)}$$

302. Suppose the reaction represented above is at equilibrium. Which of the following changes will result in an increased amount of $O_{2(g)}$?

(A) Increasing the pressure
(B) Decreasing the volume
(C) Adding more $N_{2(g)}$
(D) Decreasing the temperature
(E) Removing $NOCl_{(g)}$

$$C_{(s)} + O_{2(g)} \Leftrightarrow CO_{2(g)} + heat$$

303. Which of the following statements is true regarding the equilibrium reaction represented above?

(A) Increasing the pressure will shift the equilibrium to the right.
(B) Increasing the temperature will shift the equilibrium to the right.
(C) Adding more $C_{(s)}$ will shift the equilibrium to the right.
(D) Decreasing the temperature will result in the formation of more $O_{2(g)}$.
(E) Adding more O_2 will result in the formation of more heat.

304. In which of the following reactions at equilibrium will there be *no change* in response to a change in the volume of the reaction vessel? (Assume constant temperature.)

(A) $SO_2Cl_{2(g)} \Leftrightarrow SO_{2(g)} + Cl_{2(g)}$
(B) $H_{2(g)} + Cl_{2(g)} \Leftrightarrow 2\ HCl_{(g)}$
(C) $2\ SO_{2(g)} + O_{2(g)} \Leftrightarrow 2\ SO_{3(g)}$
(D) $4\ NH_{3(g)} + 5O_{2(g)} \Leftrightarrow 4\ NO_{(g)} + 6\ H_2O_{(g)}$
(E) $2\ N_2O_{(g)} + O_{2(g)} \Leftrightarrow 2\ NO_{(g)}$

305. A closed rigid container contains distilled water and $H_{2(g)}$ at equilibrium. Which of the following actions would increase the concentration of $H_{2(g)}$ in the water?

I. Decreasing the temperature of the water
II. Vigorous shaking
III. Injecting more $H_{2(g)}$ into the container

(A) I only
(B) II only
(C) III only
(D) I and III only
(E) I, II, and III

Questions 306–308 refer to the following chemical reaction at equilibrium.

$$2\ NH_{3(g)} \Leftrightarrow 3\ H_{2(g)} + N_{2(g)} \qquad \Delta H_{forward\ rxn} = +46.1\ kJ\ mol^{-1}$$

306. Suppose the reaction above occurs in a closed, rigid tank. After it has reached equilibrium, pure $N_{2(g)}$ is injected into the tank at a constant temperature. After it has re-established equilibrium, which of the following has a lower value compared to its value at the original equilibrium?

(A) The amount of $NH_{3(g)}$ in the tank
(B) The amount of $H_{2(g)}$ in the tank
(C) The amount of $N_{2(g)}$ in the tank
(D) K_{eq} for the reaction
(E) The total pressure in the tank

307. Which of the following changes, if occurred *alone*, would cause a decrease in the value of K_{eq} for the reaction represented above?

(A) Adding a catalyst
(B) Lowering the temperature
(C) Raising the temperature
(D) Decreasing the volume
(E) Increasing the volume

308. When the reaction represented above is at equilibrium, the ratio

$$\frac{[NH_3]^2}{[N_2][H]^3}$$

can be increased by which of the following?

I. Increasing the pressure
II. Decreasing the pressure
III. Increasing the temperature
IV. Decreasing the temperature

(A) I only
(B) I and III only
(C) I and IV only
(D) II and III only
(E) II and IV only

Questions 309–311 refer to the following reaction.

$$2\,X_{(g)} \Leftrightarrow 3\,Y_{(g)} + Z_{(g)} \qquad \Delta H_{\text{forward rxn}} > 0$$

309. The molar equilibrium concentrations for the reaction mixture represented above at 298 K are [X] = 4.0 M, [Y] = 5.0 M, and [Z] = 2.0 M. What is the value of the equilibrium constant, K_{eq}, for the reaction at 298 K?

(A) 0.06
(B) 2.50
(C) 16.0
(D) 62.5
(E) 105.5

310. Which of the following changes to the equilibrium system shown above will result in an increased quantity of Z?

I. Increasing the pressure
II. Decreasing the temperature
III. Adding more $X_{(g)}$

(A) I only
(B) II only
(C) III only
(D) II and III only
(E) I, II, and III

311. Changes to the equilibrium system above that will increase the ratio of

$$\frac{[Y]^3[Z]}{[X]^2}$$

include which of the following?

I. Decreasing the pressure
II. Increasing the temperature
III. Adding more $X_{(g)}$

(A) I only
(B) II only
(C) I and III only
(D) II and III only
(E) I, II, and III

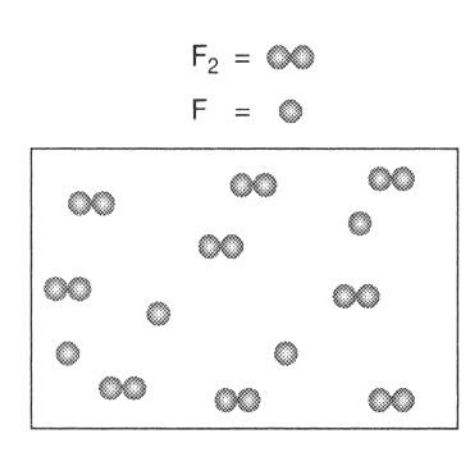

312. The diagram above represents a mixture of F_2 and F gases in a 1.0-L container at equilibrium according to the equation $F_2 \Leftrightarrow 2\,F$ at a temperature above 1,200 K. Which of the following is true of the equilibrium constant for this reaction at this temperature?

(A) $K < 0$
(B) $K = 0$
(C) $0 < K < 1$
(D) $K = 1$
(E) $K > 1$

$$2\,SO_{2(g)} + O_{2(g)} \Leftrightarrow 2\,SO_{3(g)}$$

313. At 500 K, the K_{eq} for the reaction above is 2.5×10^{10}. What is the value of K_{eq} for the *reverse* reaction at 500 K?

(A) -2.5×10^{10}
(B) 2.5×10^{-10}
(C) 2.5×10^{-11}
(D) 4.0×10^{10}
(E) 4.0×10^{-11}

$$XY_3 \Leftrightarrow X^{3+} + 3\,Y^-$$

314. A soluble compound XY_3 dissociates in water according to the equation above. If, in a 0.06-*m* solution of the compound, 30.0 percent of XY_3 dissociates, which of the following is closest to the number of moles of particles of solute per 1.0 kg of water?

(A) 0.018
(B) 0.060
(C) 0.072
(D) 0.114
(E) 0.240

315. What is the molar solubility of $BaSO_{4(s)}$ in water? (The K_{sp} for $BaSO_{4(s)}$ is 1.1×10^{-10}.)

(A) 5.5×10^{-11} M
(B) 1.10×10^{-10} M
(C) 1.05×10^{-5} M
(D) 5.5×10^{-5}
(E) 5.5×10^{-10} M

316. What is the molar solubility of $CaF_{2(s)}$ in water? (The K_{sp} for $CaF_{2(s)}$ is 4.0×10^{-11}.)

(A) $(4.0 \times 10^{-11})^{1/2}$ M
(B) $(1.0 \times 10^{-11})^{1/2}$ M
(C) 4.0×10^{-11} M
(D) $(4.0 \times 10^{-11})^{1/3}$ M
(E) $(1.0 \times 10^{-11})^{1/3}$ M

317. A saturated solution of metal hydroxide $M(OH)_2$, has a pH of 10.0 at 25°C. What is the K_{sp} value for $M(OH)_2$?

(A) 5×10^{-31}
(B) 1×10^{-20}
(C) 5×10^{-13}
(D) 2.5×10^{-9}
(E) 1×10^{-8}

318. Iron (II) hydroxide is sparingly soluble. Its solvation in water is represented by $Fe^{2+}_{(aq)} + 2\ OH^{-}_{(aq)} \Leftrightarrow Fe(OH)_{2(s)}$. True statements about the solubility of iron (II) hydroxide include:

I. Increasing the pH reduces its solubility
II. Decreasing the pH reduces its solubility
III. The value of K_{sp} for the reaction is specific to the pH of the solution in which it is dissolved.

(A) I only
(B) II only
(C) III only
(D) I and II only
(E) I, II, and III

319. A solution contains 0.2 M Ba^{2+} and 0.2 M Ca^{2+}. Which of the following CrO_4^{2-} concentrations will precipitate as much Ba^{2+} as possible without precipitating any $CaCrO_4$? (The K_{sp} of $BaCrO_4 = {\sim}1 \times 10^{-10}$ and the K_{sp} of $CaCrO_4 = {\sim}7 \times 10^{-4}$.)

(A) 3.5×10^{-3} M
(B) 7×10^{-4} M
(C) 1.5×10^{-7} M
(D) 5×10^{-10} M
(E) 7×10^{-14} M

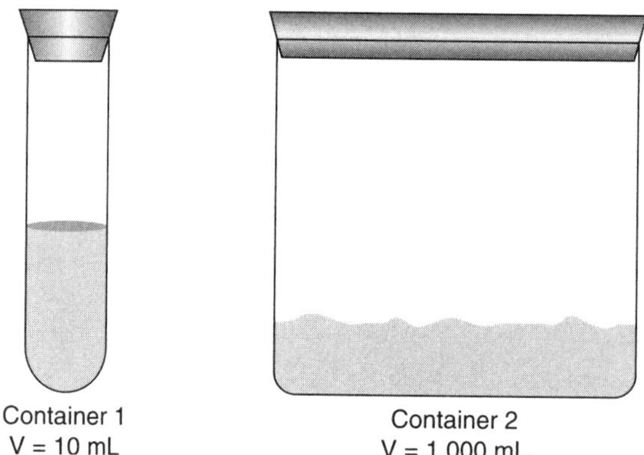

320. The figure above shows two completely sealed containers, a 10-mL tube, and a 1,000-mL bowl, each filled to a fraction of its volume with methanol. In a laboratory at 1 atm and 25°C, the vapor pressure of the methanol is

(A) lower in Container 1 because the volume of methanol is lower.
(B) higher in Container 1 because the volume of the container above the methanol is smaller.
(C) higher in Container 2 because the surface area of the methanol is larger.
(D) higher in Container 2 because the volume of the container above the methanol is greater.
(E) equal in Containers 1 and 2 because they are at the same temperature.

$$\text{liquid} + \text{heat} \Leftrightarrow \text{gas}$$

321. The vaporization of a liquid is represented in the equation above. Which of the following most accurately accounts for the fact that the vapor pressure of liquids increases with increasing temperature?

(A) Vaporization is exothermic.
(B) A liquid at a particular temperature continues to vaporize until it has completely evaporated.
(C) If a chemical system at equilibrium experiences a change in temperature, the equilibrium shifts to counteract the change.
(D) The condensation of gases takes longer than the vaporization of liquids.
(E) The condensation of gases releases more energy than is absorbed by vaporizing liquids.

322. Which of the following best accounts for the fact that, at the same altitude, the partial pressure of water vapor in the atmosphere can increase significantly with increasing temperature, but the partial pressures of N_2, O_2, Ar, and CO_2 stay relatively constant?

(A) The vapor pressures of N_2, O_2, Ar, and CO_2 are much greater than that of H_2O.
(B) The water cycle (evaporation, condensation, precipitation) causes water to constantly change between liquid and gas phases.
(C) The conditions required for atmospheric N_2, O_2, Ar, and CO_2 to be in equilibrium with their liquid phases do not exist on the earth's surface.
(D) Water in the atmosphere constantly forms water droplets around condensation nuclei whereas N_2, O_2, Ar, and CO_2 only condense in or above the stratosphere.
(E) Water vapor is constantly being added to the atmosphere by cellular respiration and combustion.

$$Z_{(s)} \Leftrightarrow Z_{(l)}$$

323. Consider a closed, adiabatic system consisting of a mixture of liquid and solid substance Z at equilibrium at its melting point. Which of the following statements is true regarding the system?

(A) The entropy of the system is at a maximum.
(B) The entropy of the system is at a minimum.
(C) The entropy of the system will increase over time.
(D) The entropy of the system is zero.
(E) The entropy of pure substances does not change if at a constant temperature.

Questions 324–328 refer to the information in the table below.

Acid	K_a at 298 K
HOCl	2.9×10^{-8}
HOBr	2.4×10^{-9}
HOI	??

324. Which of the following is true regarding the strengths of HOCl and HOBr?

(A) HOBr is a stronger acid because Br is a larger atom than Cl, and loses its proton more easily than HOCl.
(B) HOCl is a stronger acid because Cl is more electronegative than HOBr.
(C) The strengths of HOCl and HOBr are dependent on the pH of their respective solutions.
(D) The strengths of HOCl and HOBr are dependent on their respective concentrations.
(E) HOBr is the stronger acid because the negative log of its K_a, the pK_a, is larger than that of HOCl.

325. Which of the following statements is the most accurate prediction about the strength of acid HOI?

(A) HOI is the strongest acid because I is a larger atom than Br and Cl, and therefore the proton dissociates more readily in solution.
(B) HOI is the strongest acid because the O–H bond in HOI is stronger than the OH bond in HOCl or HOBr.
(C) HOI is the weakest acid because its conjugate base, OI–, is less stable than the conjugate bases of HOCl and HOBr.
(D) The strength of HOI is dependent on the concentrations and pH of its solution.
(E) HOI is the stronger acid than HOBr and HOCl because the negative log of its K_a, the pK_a, is larger than the pK_a of HOBr and HOCl.

326. Which of the following is the correct name for HOCl?

(A) Hydrochloric acid
(B) Hypochloric acid
(C) Hypochorous acid
(D) Perchlorous acid
(E) Hydrogen chloride

$$OCl^- + H_2O \Leftrightarrow HOCl + OH^-$$

327. The hydrolysis of KOCl is represented above. Which of the following is the correct set up for determining the equilibrium constant of the reaction?

(A) $2.9 \times 10^{-8} = [OH^-][HOCl]/[OCl^-]$
(B) $2.9 \times 10^{-8} = [OH^-][HOCl]/[OCl^-][H_2O]$
(C) $1 \times 10^{-14} = [OH^-][2.9 \times 10^{-8}]/[OCl^-]$
(D) $K_b = 1.0 \times 10^{-14}/2.9 \times 10^{-8}$
(E) $K_b = 2.9 \times 10^{-8}/1.0 \times 10^{-14}$

328. Which of the following shows the correct way to calculate the [OH⁻] concentration in a 1.5-M solution of KOCl at 298 K? Assume the $K_{hydrolysis} = 3.4 \times 10^{-7}$.

(A)

	$[OCl^-]$	$[H_2O]$	$[HOCl]$	$[OH^-]$	
[Initial]	1.5	1.0	0	3.4×10^{-7}	
Change	$-x$	$-x$	x	x	$1 \times 10^{-14} = x^2/1.5$
[Equilibrium]	$1.5 - x$	$1.0 - x$	x	$3.4 \times 10^{-7} + x$	$x = [OH^-]$

(B)

	$[OCl^-]$	$[H_2O]$	$[HOCl]$	$[OH^-]$	
[Initial]	1.5	1.5	0	~0	
Change	$-x$	$-x$	x	x	$3.4 \times 10^{-7} = x^2/1.5$
[Equilibrium]	$1.5 - x$	$1.5 - x$	x	x	$x = [OH^-]$

(C)

	$[OCl^-]$	$[HOCl]$	$[OH^-]$	
[Initial]	1.5	0	1×10^{-14}	
Change	$-x$	x	x	$3.4 \times 10^{-7} = x^2/1.5$
[Equilibrium]	$1.5 - x$	x	$1 \times 10^{-14} + x$	$x = [OH^-]$

(D)

	$[OCl^-]$	$[HOCl]$	$[OH^-]$	
[Initial]	1.5	0	~0	
Change	$-x$	x	x	$3.4 \times 10^{-7} = x^2/1.5$
[Equilibrium]	$1.5 - x$	x	x	$x = [OH^-]$

(E)

	$[OCl^-]$	$[HOCl]$	$[OH^-]$	
[Initial]	1.5	0	3.4×10^{-7}	
Change	$-x$	x	x	$1 \times 10^{-14} = x^2/1.5$
[Equilibrium]	$1.5 - x$	x	$3.4 \times 10^{-7} + x$	$x = [OH^-]$

Questions 329 and 330 refer to the reaction A ⇔ B. The concentrations of A and B throughout the reaction are shown in the figure below.

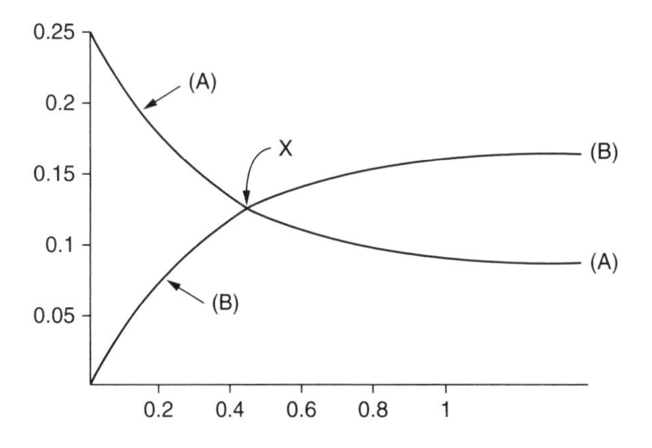

329. Which of the following statements is true regarding the progress of the reaction?

(A) At point X, the rate of A→B equals the rate B→A.
(B) From 0.0 to 0.8 sec, the concentration of Reactant A is increasing.
(C) From 0.0 to 0.8 sec, the rate of A→B exceeds the rate of B→A.
(D) After 1 sec, the reaction has gone to completion.
(E) If the reaction was observed after 10 minutes, the concentrations of A and B would be equal.

330. Suppose at time 1.5 sec, more B was added to the system (enough to momentarily raise the concentration to 0.25 M). In which of the following ways is the system expected to respond?

(A) The forward reaction would be pushed to completion, consuming all of Reactant A.
(B) The equilibrium state would be overwhelmed and the excess B would precipitate.
(C) The equilibrium state would be overwhelmed forcing the reverse reaction to go to completion.
(D) The concentration of A would increase.
(E) The equilibrium would adjust to consume the excess B and produce a new ratio of [A]/[B].

Acid–Base Chemistry

Questions 331–334 refer to the following answer choices:

(A) Proton acceptor
(B) Proton donor
(C) Electron pair acceptor
(D) Electron pair donor
(E) Hydroxide generator

331. Bronsted–Lowry acid

332. Lewis base

333. Arrhenius base

334. Forms coordinate covalent bonds with acids

Questions 335–339 refer to aqueous solutions containing 1:1 mole ratios of the following pairs of substances. Assume all concentrations are 1 M.

(A) NH_3 and H_3CCOOH (acetic acid)
(B) KOH and NH_3
(C) HCl and KCl
(D) H_3PO_4 and KH_2PO_4
(E) NH_3 and NH_4Cl

335. The solution with the highest pH

336. The solution with the lowest pH

337. The solution with the pH closest to neutral

338. A buffer at an alkaline pH

339. A buffer at an acidic pH

340. What is the H^+ concentration of a 0.02-M nitrous acid (HNO_2) solution? (The K_a for HNO_2 is 4.5×10^{-4}.)

(A) 2.25×10^{-2}
(B) 3.0×10^{-3}
(C) 5.1×10^{-4}
(D) 9.0×10^{-6}
(E) 2.6×10^{-7}

$$F^-_{(aq)} + H_2O_{(l)} \Leftrightarrow HF_{(aq)} + OH^-_{(aq)}$$

341. Which of the following statements is true of the reaction represented above?

(A) H_2O is the conjugate acid of F^-.
(B) OH^- is the conjugate acid of H_2O.
(C) HF is the conjugate base of F^-.
(D) HF and H_2O are conjugate acid–base pairs.
(E) HF and H_2O are both Bronsted–Lowry acids.

$$NH_3 + H_2O \longleftrightarrow NH_4^+ + OH^-$$

342. In the reaction above, H_2O acts as:

(A) An acid
(B) A base
(C) A conjugate acid
(D) An oxidizing agent
(E) A reducing agent

343. All of the following increase the strength of an oxyacid *except*:

(A) A strongly electronegative central atom
(B) Electronegative atoms bonded to the central atom
(C) Electronegative atoms bonded to atoms other than the central atom
(D) An increased number of oxygen atoms bonded to the central atom
(E) An increased number of hydrogens

344. It is possible to create all of the following solutions by mixing 0.25 M and 0.35 M HCl *except*:

(A) 0.34 M HCl
(B) 0.31 M HCl
(C) 0.29 M HCl
(D) 0.26 M HCl
(E) 0.24 M HCl

345. The acid dissociation constant for a weak monoprotic acid HA is 5.0×10^{-9}. The pH of a 0.5 M HA solution is closest to:

(A) 2
(B) 3
(C) 4
(D) 5
(E) 6

$$HX_{(aq)} + Y^-_{(aq)} \Leftrightarrow X^-_{(aq)} + HY_{(aq)}$$

346. The reaction above represents a general equilibrium between monoprotic acids HX and HY. If the equilibrium constant for this reaction was 2.5×10^3, which of the following could be correctly concluded about the chemical species involved?

(A) HY is a stronger acid than HX.
(B) Y^- is a stronger base than X^-.
(C) HY is the conjugate base of Y^-.
(D) X^- is the conjugate acid of HX.
(E) The pH of a solution containing a 1:1 mole ratio of HX and Y^- is 7.

Questions 347–349 refer to the following titration.

A solution of a weak monoprotic acid is titrated with a 0.1-M strong base, NaOH. The titration curve is shown below.

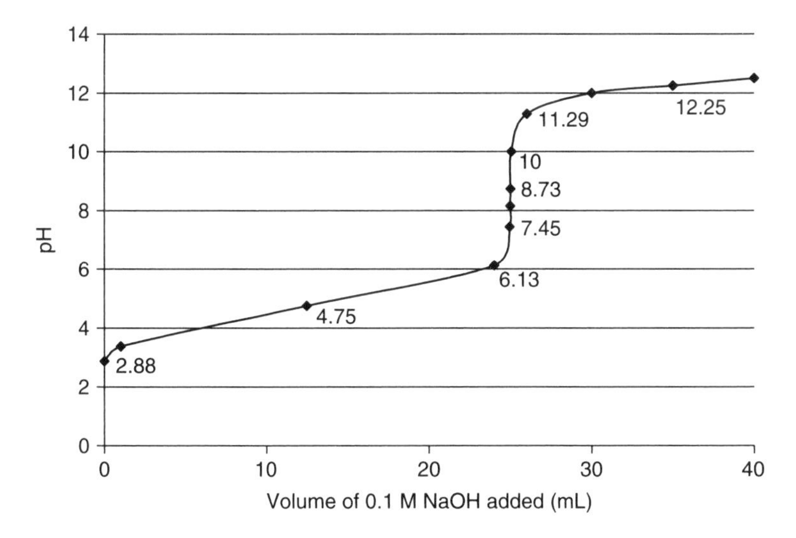

347. From the curve, what is the value of the pH where the number of moles of strong base added is equal to the number of moles of weak acid in the initial solution?

(A) 4.75
(B) 7.00
(C) 7.45
(D) 8.73
(E) 11.3

348. At which pH are the concentrations of the weak acid and its conjugate base approximately equal?

(A) 2.88
(B) 4.75
(C) 7.00
(D) 8.73
(E) 13.0

349. The buffer region of this titration curve can be found between which of the following pH values?

(A) 2.88 and 6.13
(B) 4.75 and 8.73
(C) 6.13 and 11.29
(D) 8.73 and 11.29
(E) 11.29 and 12.25

350. Which of the following solutions at a concentration of 0.1 M has a pH > 7?

(A) NaCl
(B) KI
(C) $HC_2H_3O_2$
(D) LiF
(E) NaBr

351. Which of the following is the conjugate acid of NH_3?

(A) H^+
(B) N_2
(C) NH_2^-
(D) NH_3^+
(E) NH_4^+

352. At 25°C, aqueous solutions with a pH of 6 have a hydroxide ion concentration, $[OH^-]$, of:

(A) 1×10^{-6} M
(B) 1×10^{-8} M
(C) 0.006 M
(D) 6 M
(E) 8 M

353. Which of the following steps would convert 100 mL of KOH solution with a pH of 13 to a KOH solution with a pH of 12?

(A) Dilute the solution by adding 10 mL of distilled water.
(B) Dilute the solution by adding 900 mL of distilled water.
(C) Dilute the solution with an equal amount of KOH solution with a pH of 1.
(D) Add 10 mL of 1 M HCl solution.
(E) Add 100 mL of 1 M HCl solution.

354. Citric acid ($H_3C_6H_5O_7$) is a triprotic acid with $K_1 = 8.4 \times 10^{-4}$, $K_2 = 1.8 \times 10^{-5}$, and $K_3 = 4.0 \times 10^{-6}$. In a 0.01-M aqueous solution of citric acid, which of the following species is present in the *lowest* concentration?

(A) $H_3O^+{}_{(l)}$
(B) $H_3C_6H_5O_{7(aq)}$
(C) $H_2C_6H_5O_7{}^-{}_{(aq)}$
(D) $H_1C_6H_5O_7{}^{2-}{}_{(aq)}$
(E) $C_6H_5O_7{}^{3-}{}_{(aq)}$

355. Mixtures that would be useful as buffers include which of the following?

I. NH_3 and NH_4Cl
II. HCl and KCl
III. HF and KF

(A) I only
(B) I and II only
(C) I and III only
(D) II and III only
(E) I, II, and III

356. A sample of 200 mL of 0.20 M $Sr(OH)_2$ is added to 800 mL of 0.80 M $Ba(OH)_2$. Which of the following best approximates the pH of the final solution?

(A) 0
(B) 1
(C) 12
(D) 13
(E) 14

357. The pH of a solution prepared by the addition of 100 mL 0.002 M HCl to 100-mL distilled water is closest to:

(A) 1.0
(B) 1.5
(C) 2.0
(D) 3.0
(E) 4.0

358. The pH of a solution prepared by the addition of 5.0 mL 0.02 M $Ba(OH)_2$ to 15-mL distilled water is closest to:

(A) 2.0
(B) 4.0
(C) 11.0
(D) 12.0
(E) 13.0

Questions 359–361 refer to the figures below. The figures show a burette used in the titration of an acid with a 0.5-M solution of $Sr(OH)_2$. Figure 1 shows the level of $Sr(OH)_2$ at the start of the titration and Figure 2 shows the level of $Sr(OH)_2$ at the end of the titration. The indicator used in this experiment was phenophthalein.

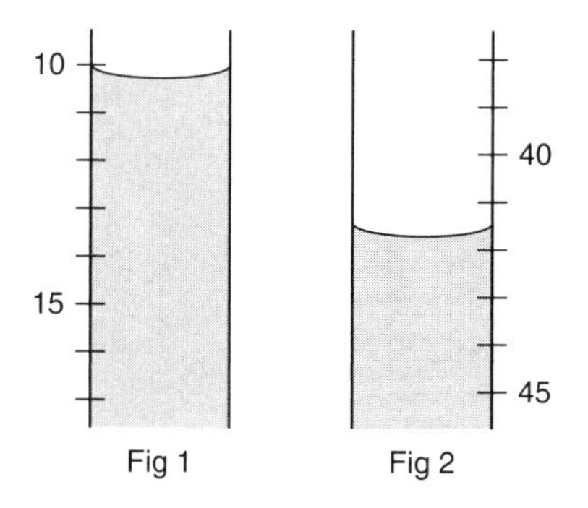

359. What evidence indicates that the endpoint of the titration has been reached?

(A) The color of the solution in the burette changes from colorless to pink.
(B) The color of the solution in the burette changes from pink to colorless.
(C) The color of the solution in the flask below changes from colorless to pink.
(D) The color of the solution in the flask below changes from pink to colorless.
(E) The pH of the solution does not change.

360. The volume of $Sr(OH)_2$ used to neutralize the acid was closest to:

(A) 3.5 mL
(B) 10.60 mL
(C) 31.80 mL
(D) 42.04 mL
(E) 42.40 mL

361. In a different titration, 25 mL of the same 0.5 M $Sr(OH)_2$ solution was used to titrate 35 mL HCl solution to the equivalence point. What is the molarity of the HCl solution?

(A) 0.5 M
(B) 0.7 M
(C) 1.4 M
(D) 2.1 M
(E) 2.8 M

Electrochemistry

362. The oxidation state most common to ions of Fe, Mn, and Zn in solution is:

(A) +1
(B) +2
(C) +3
(D) +4
(E) +5

363. The oxidation state of chlorine in $HClO_4$ is:

(A) −1
(B) +1
(C) +3
(D) +5
(E) +7

364. In which of the following species does chromium have the same oxidation number as in $Cr_2O_4^{2-}$?

(A) $Cr_2(O_2CCH_3)_4$
(B) $Cr(CO)_6$
(C) $Cr_2O_7^{2-}$
(D) $K_3[Cr(O_2)4]$
(E) Cr_2O_3

365. A 0.1-M solution of H_3PO_4 is a better conductor of electricity than a 0.1-M solution of NaCl. Which of the following *best* explains this observation?

(A) NaCl is less soluble than H_3PO_4.
(B) NaCl has a lower molar mass than H_3PO_4.
(C) H_3PO_4 dissociates to produce particles of a larger size than NaCl.
(D) H_3PO_4 is weakly acidic, which increases the flow of charge through a solution.
(E) Fewer moles of ions are present in the NaCl solution than in the same volume of an Na_3PO_4 solution.

$$Cl_{2(g)} + 2\ Br^-_{(aq)} \rightarrow 2\ Cl^-_{(aq)} + Br_{2(aq)}$$

366. Which of the following choices *best* explains why the reaction represented above generates electricity in a galvanic cell?

(A) Cl_2 is a stronger oxidizing agent than Br_2.
(B) Cl_2 loses electrons more easily than Br_2.
(C) Br atoms have more electrons than Cl atoms.
(D) Br_2 is more stable than Cl_2.
(E) Cl^- is more stable than Br^-.

$$3\ Cu_{(s)} + 8\ H^+_{(aq)} + 2\ NO_3^-{}_{(aq)} \rightarrow 3\ Cu^{2+}_{(aq)} + 2\ NO_{(g)} + 4\ H_2O_{(l)}$$

367. Which of the following statements is/are true regarding the reaction above?

I. $Cu_{(s)}$ acts as an oxidizing agent.
II. $H^+_{(aq)}$ gets reduced.
III. The oxidation state of nitrogen changes from +5 to +2.

(A) I only
(B) II only
(C) III only
(D) II and III only
(E) I, II, and III

Questions 368 and 369 refer to a galvanic cell constructed of two half-cells and the two half-reactions represented below.

$$Zn^{2+}_{(aq)} + 2\ e^- \rightarrow Zn_{(s)} \qquad E° = -0.76\ V$$

$$Ag^+_{(aq)} + e^- \rightarrow Ag_{(s)} \qquad E° = 0.80\ V$$

368. As the cell operates, which of the following species is contained in the half-cell containing the cathode?

I. Zn^{2+}
II. Zn^+
III. Ag^+

(A) I only
(B) II only
(C) III only
(D) I and II only
(E) I and III only

369. What is the standard cell potential for this galvanic cell?

(A) –0.04 V
(B) 0.04 V
(C) 0.84 V
(D) 1.56 V
(E) 2.36 V

$$2\ KMnO_4 \rightarrow K_2MnO_4 + MnO_{2(s)} + O_2$$

370. The photodecomposition of permanganate is shown above. As this reaction proceeds in the forward direction and the reaction species are considered from left to right, the oxidation number of Mn changes from:

(A) +1 to +2 and +2
(B) +5 to +2 and +2
(C) +5 to +6 and +4
(D) +7 to +6 and +4
(E) +7 to +2 and +2

Questions 371 and 372 refer to an electrolytic cell involving the following reaction.

$$2\ Al_2O_{3(s)} \rightarrow 4\ Al_{(l)} + 3\ O_{2(g)}$$

371. Which of the following processes occurs in this reaction?

(A) O_3 acts as an oxidizing agent.
(B) O_3^{6-} acts as an oxidizing agent.
(C) Al^{3+} is reduced at the cathode.
(D) Al is oxidized at the anode.
(E) Aluminum is converted from a negative to a neutral oxidation state.

372. A steady current of 15 amperes is passed through the cell for 20 minutes. Which of the following correctly expresses how to calculate the number of grams of aluminum produced by this cell (1 faraday = 96,500 coulombs)?

(A) $\dfrac{(27)(3)(60)}{(15)(20)(96{,}500)}$

(B) $\dfrac{(3)(96{,}500)}{(15)(20)(60)(27)}$

(C) $\dfrac{(15)(20)(27)}{(3)(96{,}500)}$

(D) $\dfrac{(15)(20)(60)(27)}{(3)(96{,}500)}$

(E) $\dfrac{(15)(20)(27)(3)}{(96{,}500)(60)}$

$X_{(s)} + 2\,Ag^{+}_{(aq)} \rightarrow 2\,Ag_{(s)} + X^{2+}_{(aq)}$ $E° = +2.27$

$2\,Ag^{+}_{(aq)} + 2\,e^{-} \rightarrow 2\,Ag_{(s)}$ $E° = +0.80$ V

373. According to the information above, what is the standard reduction potential for the half-reaction $X^{2+}_{(aq)} + 2e^{-} \rightarrow X_{(s)}$?

(A) +0.67 V
(B) –0.67 V
(C) +1.47 V
(D) –1.47 V
(E) +3.07 V

374. An electric current of 1.00 ampere is passed through an aqueous solution of $FeCl_3$. Assuming 100 percent efficiency, how long will it take to electroplate exactly 1.00 mole of iron metal?

(1 faraday = 96,500 coulombs = 6.02×10^{23} electrons)

(A) 32,200 sec
(B) 96,500 sec
(C) 193,000 sec
(D) 289,500 sec
(E) 386,000 sec

Questions 375–381 refer to a galvanic cell made of two half-cells. In one half-cell, a Zn electrode is bathed in a 1.0 M $ZnSO_4$ solution. The other cell contains a 1 M HCl solution and a hydrogen electrode.

$Zn^{2+} + 2\,e^{-} \rightarrow Zn_{(s)} E° = -0.76$ V

$2\,H^{+} + 2\,e^{-} \rightarrow H_{2(g)} E° = 0.00$ V

375. Which of the following is the correct electron configuration for Zn^{2+}?

(A) $1s^2\,2s^2\,2p^6\,3s^2\,3p^6\,4s^2$
(B) $1s^2\,2s^2\,2p^6\,3s^2\,3p^6\,3d^{10}$
(C) $1s^2\,2s^2\,2p^6\,3s^2\,3p^6\,3d^{10}\,4s^2$
(D) $1s^2\,2s^2\,2p^6\,3s^2\,3p^6\,3d^8\,4s^2$
(E) $1s^2\,2s^2\,2p^6\,3s^2\,3p^6\,3d^9\,4s^1$

376. Which of the following correctly identifies and justifies which species, Zn or Zn^{2+}, has the highest ionization energy?

(A) Zn, electrons have a greater affinity for pure metals.
(B) Zn, it has a greater electron to proton ratio.
(C) Zn^{2+}, it has a larger radius.
(D) Zn^{2+}, it has a greater effective nuclear charge.
(E) Zn^{2+}, it has a greater electron-to-proton ratio.

377. Processes that occur at the anode include which of the following?

I. $Zn_{(s)}$ is oxidized to $Zn^{2+}_{(aq)}$.
II. $Zn^{2+}_{(aq)}$ is reduced to $Zn_{(s)}$.
III. $H_{2(g)}$ is oxidized to $H^{+}_{(aq)}$.
IV. 2 $H^{+}_{(aq)}$ is reduced to $H_{2(g)}$.

(A) I only
(B) II only
(C) III only
(D) I and III only
(E) II and IV only

378. The salt bridge is filled with a saturated solution of KNO_3. Processes that occur at the salt bridge include which of the following?

I. K^+ moves into the anode half-cell.
II. K^+ moves into the cathode half-cell.
III. NO_3^- moves into the anode half-cell.
IV. NO_3^- moves into the cathode half-cell.

(A) I only
(B) III only
(C) I and IV only
(D) II and III only
(E) I, II, III, and IV

379. Which of the following is true regarding the HCl half-cell?

(A) The reduction of H^+ to H_2 is neither spontaneous nor non-spontaneous.
(B) The reduction of H^+ and oxidation of H_2 are at equilibrium.
(C) All electrochemical cells use a hydrogen half-cell.
(D) It is used to measure the reduction potential of Zn^{2+} because the reduction potential has a measured value of zero volts.
(E) The E° value of 0 V for the hydrogen half-cell is arbitrary.

380. Which of the following is true regarding the standard free energy change (ΔG°) of the cell?

(A) It is negative because the standard reduction potential of Zn^{2+} is negative.
(B) It is negative because the sum of the standard reduction potentials of Zn^{2+} and H^+ is –0.76 V.
(C) It is negative because the oxidation of Zn is spontaneous.
(D) It is positive because the reduction of hydrogen ions is spontaneous.
(E) It has a value of zero because the reaction is at equilibrium.

381. Which of the following changes to the voltage, if any, would result from increasing the concentration of $ZnSO_4$?

(A) No change, the voltage of the cell is determined only by the standard reduction potentials of the electrodes.
(B) Increased voltage, the value of Q will fall below 1 and the log term in the Nernst equation will become more negative.
(C) Increased voltage, the reaction becomes more favorable as the concentration of Zn^+ increases.
(D) Decreased voltage, the value of Q will fall below 1 and the log term in the Nernst equation will become more negative.
(E) Decreased voltage, the increased concentration of Zn^{2+} will inhibit the oxidation of $Zn_{(s)}$.

382. Iron nails are often electroplated (galvanized) with zinc. Which of the following is true regarding this process (E° of $Zn^{2+} \rightarrow Zn_{(s)} = -0.76$ V, E° of $Fe^{2+} \rightarrow Fe_{(s)} = -0.44$ V)?

(A) Zn serves as a sacrificial anode for Fe.
(B) Zn serves as an airtight seal preventing the electrons from iron from reacting with oxygen.
(C) The galvanization process produces an alloy of Zn and Fe that is stronger and more resistant to oxidation than Fe alone.
(D) The galvanization process is spontaneous, thereby reducing the energy cost of producing nails.
(E) The oxidation of Zn allows the reduction of Fe^{2+} to proceed spontaneously.

383. Which of the following correctly relates the quantities needed for the conversion of chemical energy into electrical energy in a galvanic cell?

(A) The sum of the E° cell of the two half-cells.
(B) The product of the cell voltage and the total charge passed through the cell.
(C) The product of cell voltage and the natural log (ln) of K, the equilibrium constant for the redox reaction.
(D) The product of the gas constant and the voltage divided by the total charge passed through the cell.
(E) The difference between the voltages of the two half-cells.

384. All of the following statements are consistent with the operation of a galvanic cell *except*:

(A) During cell operation, reactant concentration and product formation decreases.
(B) The electrons that flow from anode to cathode are generated at the anode.
(C) E° values are specific for a particular reaction and do not vary with reactant concentration.
(D) At equilibrium, there is no net transfer of electrons and the cell cannot do work.
(E) As the product-to-reactant ratio increases, the work that can be done by the cell decreases.

Questions 385–387 refer to the following redox reaction that occurs in a lithium ion battery.

$$Li_{(s)} + CoO_2 \rightarrow Li^+ + LiCoO_{2(s)} E_{cell} = 3.4 \text{ V}$$

385. Which of the following reactions occurs at the anode?

(A) $Li_{(s)} + e^- \rightarrow Li^+$
(B) $Li_{(s)} \rightarrow Li^+ + e^-$
(C) $Li^+ + e^- \rightarrow Li_{(s)}$
(D) $CoO_2 + e^- \rightarrow CoO^{2-}$
(E) $CoO_2 \rightarrow CoO^{2+}$

386. All of the following are advantages of a lithium-ion battery *except:*

(A) Li^+ has the most negative standard reduction potential.
(B) Fewer than 7-g Li are needed to provide 1-mol electron to the battery.
(C) Lithium is a powerful reducing agent.
(D) Lithium-ion batteries can be charged hundreds of times without deterioration.
(E) Lithium is a highly reactive metal that requires a nonaqueous electrolyte solution.

387. Features common to both galvanic and electrolytic cells include which of the following?

I. Oxidation at the anode
II. Can perform electrolysis
III. Spontaneous

(A) I only
(B) II only
(C) III only
(D) I and II only
(E) I, II, and III

Nuclear Chemistry

$$^{235}_{92}U + ^{1}_{0}n \rightarrow ^{141}_{55}Cs + 3\ ^{1}_{0}n + X$$

388. The reaction above represents the bombardment of uranium with a neutron. Which of the following nuclides is represented by X?

(A) $^{141}_{37}Rb$
(B) $^{95}_{37}Rb$
(C) $^{94}_{37}Rb$
(D) $^{93}_{37}Rb$
(E) $^{92}_{37}Rb$

$$^{27}_{13}Al + ^{4}_{2}\alpha \rightarrow ^{1}_{0}n + X$$

389. The reaction above shows the synthesis of element X. Which of the following nuclides is represented by X?

(A) $^{31}_{15}P$
(B) $^{30}_{15}P$
(C) $^{28}_{15}P$
(D) $^{28}_{13}Al$
(E) $^{31}_{13}Al$

$$X + ^{4}_{2}\alpha \rightarrow ^{17}_{8}O + ^{1}_{1}p$$

390. The alpha decay represented above produces $^{17}_{8}O$ and a proton. Which of the following nuclides is represented by X?

(A) $^{18}_{7}N$
(B) $^{14}_{7}N$
(C) $^{13}_{7}N$
(D) $^{18}_{9}F$
(E) $^{14}_{9}F$

Questions 391–395 refer to the following types of radioactive decay.

(A) $^{4}_{2}He$ ($^{4}_{2}\alpha$)
(B) $^{0}_{-1}e$ ($^{0}_{-1}\beta$)
(C) $^{0}_{1}e$
(D) gamma radiation (γ)
(E) $^{1}_{0}n$

391. Has no mass

392. Least penetrative

393. Most penetrative

394. Carries the greatest positive charge

395. Increases the atomic number without changing the mass number

Questions 396 and 397 refer to the following graph of binding energy per nucleon.

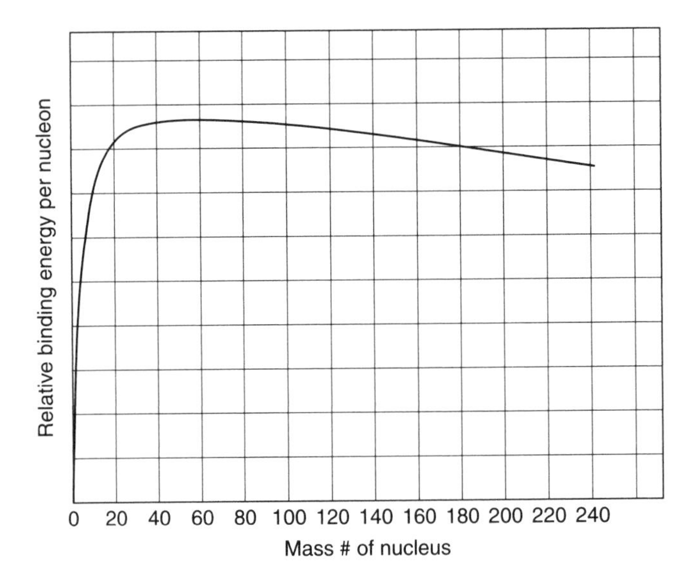

396. According to the graph, which of the following atoms has the most stable nucleus?

(A) H
(B) He
(C) Fe
(D) Au
(E) U

397. All of the following are true statements regarding the binding energy of a nuclide *except*:

(A) The nucleus with the highest binding energy means it is the nucleus with the most attractive forces.
(B) Energy per nucleon compares nuclides on a common basis.
(C) The binding energy is a function of attractive versus repulsive forces in the nucleus.
(D) The magnitude of the attractive forces within the nucleus is greater than the magnitude of like-charge repulsion in the nucleus.
(E) The total system has less potential energy than the sum of its parts.

398. The difference between the mass of an atom and the sum of the masses of its protons, neutrons, and electrons is called:

(A) Nuclear decay
(B) Nuclear-binding energy
(C) Mass defect
(D) Mass-energy equivalence
(E) Transmutation

399. Which of the following correctly describes what happens to the mass that is lost when a nucleus is formed?

(A) It is emitted from the nucleus as an α (alpha) particle.
(B) It is emitted from the nucleus in the form of a β (beta) particle.
(C) It is emitted as a neutrino.
(D) It is emitted as energy.
(E) What happens to the lost mass is not currently understood.

400. True statements regarding the alkali and naturally occurring lanthanide metals include:

I. The chemical reactivity of the alkali metals is due to their low first ionization energies.
II. Radioactive nuclei can only become more stable by forming compounds.
III. The alkali and lanthanide metals increase their chemical stability by forming compounds.

(A) I only
(B) II only
(C) III only
(D) I and II only
(E) I and III only

401. Coulomb's law states that like charges repel. Which of the following most accurately explains how the protons in the nucleus can form a stable nucleus?

(A) The positive charges attract each other when they are within a very short distance of each other.
(B) There are negative charges in the nucleus that neutralize the positive charges.
(C) The neutrons cancel out the charges of the protons.
(D) The repulsive force is very small because protons are spaced far from other protons.
(E) The neutrons form a tight cage around the protons that keep them from leaving the nucleus.

402. The average density of an atomic nucleus is on the order of magnitude closest to which of the following? (The density of osmium, Os, the heaviest known element, is 23 g cm^{-3}. The mass of an average Zn nucleus is approximately 1×10^{-22} g and the radius of a nucleus of that mass is approximately 5×10^{-13} cm.)

(A) 10^{-14} g cm^{-3}
(B) 10^{3} g cm^{-3}
(C) 10^{14} g cm^{-3}
(D) 10^{28} g cm^{-3}
(E) 10^{63} g cm^{-3}

403. In 325 days, a 30-gram sample of ^{95}Zr decayed until approximately 1 gram of ^{95}Zr remained in the sample. The half-life of ^{95}Zr is closest to:

(A) 11 days
(B) 29 days
(C) 65 days
(D) 81 days
(E) 162 days

404. If 93.75 percent of a sample of pure radioisotope X decays in 24 days, what is the half-life of X?

(A) 4 days
(B) 6 days
(C) 12 days
(D) 18 days
(E) 24 days

405. If 12.5 percent of a sample of pure radioisotope Z remains after 30 days, what is the half-life of Z?

(A) 2.4 days
(B) 8 days
(C) 10 days
(D) 12.5 days
(E) 30 days

406. The half-life of radioisotope J is 2 years. If the initial amount of J present is 60 grams, approximately how much is expected to remain after 12 years?

(A) 30 grams
(B) 12 grams
(C) 5 grams
(D) 2 grams
(E) 1 gram

407. Which of the following particles is emitted by an atom of ^{40}K when it decays into an atom of ^{40}Ar?

(A) Electron (β^-)
(B) Positron (β^+)
(C) Alpha ($^4{}_2He$)
(D) Gamma photon (γ)
(E) Neutron ($^1{}_0n$)

408. Which of the following graphs of the concentration of radioisotope (RI) remaining versus time is consistent with radioactive decay kinetics?

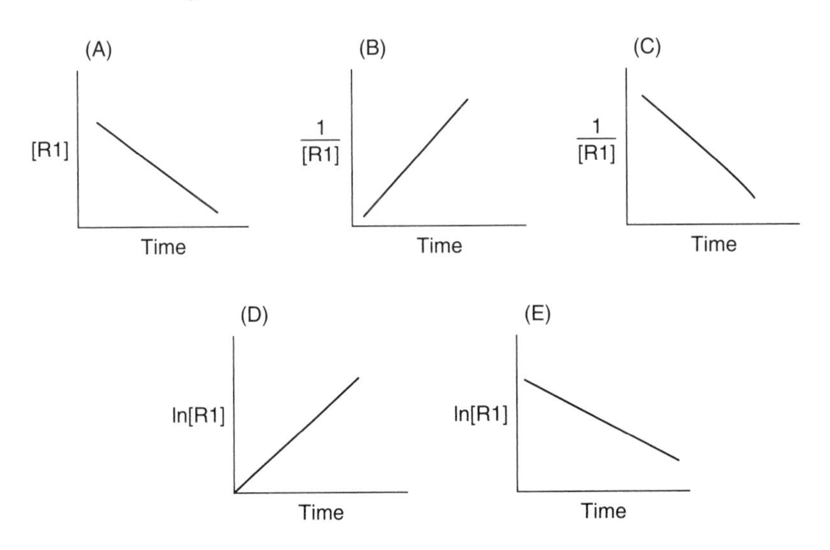

Descriptive

409. Which of the following elements forms a covalent network solid by combining with oxygen?

(A) N
(B) O
(C) S
(D) Si
(E) P

Questions 410–415 refer to the following solid compounds.

(A) $KMnO_4$
(B) $CuSO_4$
(C) NaCl
(D) $FeCl_3$
(E) NH_4NO_3

410. The white, soluble salt responsible for the salinity of the ocean

411. Produces a blue solution when mixed with water

412. Soluble, yellow solid

413. A purple solid that produces a purple solution when mixed with water

414. A deliquescent compound that turns to blue when hydrated

415. A soluble white solid used as a fertilizer

Questions 416–420 refer to the following compounds:

(A) C_3H_8
(B) HF
(C) H_2O_2
(D) H_2S
(E) CCl_3F

416. A gaseous fuel used for heating and in gas-burning barbeque grills

417. A refrigerant linked to the thinning of the ozone layer in the stratosphere

418. In solution, this compound has been used as disinfectant for minor skin wounds

419. A colorless gas with a foul odor

420. A weak acid that cannot be stored in glass containers

421. Which of the following nonmetals is a good conductor of electricity in the solid form?

(A) I_2
(B) S_6
(C) $C_{(graphite)}$
(D) $C_{(diamond)}$
(E) $P_{4(white)}$

422. Naturally occurring amino acids contain all of the following elements *except*:

(A) Carbon
(B) Nitrogen
(C) Oxygen
(D) Hydrogen
(E) Chlorine

423. All of the following oxides exist as a gas at 25°C and 1 atm *except*:

(A) NO
(B) N_2O
(C) CO_2
(D) SiO_2
(E) SO_2

Questions 424–428 refer to the following organic compounds.

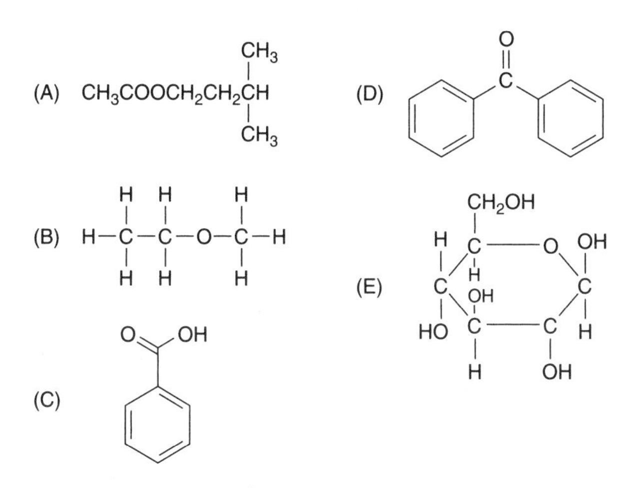

424. A ketone

425. Very water soluble

426. An ester with the scent of bananas

427. Benzoic acid

428. An ether

Laboratory Procedure

429. Which of the following is an acceptable laboratory practice?

(A) Placing hot objects on a balance
(B) Diluting a solution in a volumetric flask with hot water
(C) Using 5 mL of phenolphthalein to titrate 20 mL of an acidic solution
(D) Adding water slowly to a preweighed solid acid in a dry Erlenmeyer flask
(E) Rinsing a burette with a standardized solution before filling it with the standardized solution

430. The proper procedure for the dilution of concentrated NaOH is to slowly add the base to a beaker of water rather than slowly adding water to the beaker of base. This precaution is to ensure that

(A) there is sufficient time for the full ionization of NaOH.
(B) the water does not float on top of the denser base and remain unmixed.
(C) the base does not react with the dry beaker or impurities in the beaker.
(D) there is a enough water to absorb the heat released.
(E) there is enough water for the OH^- ions to be completely dissolved.

431. A student accidentally splashes a concentrated, strong acid onto his bare skin. Which of the following is the safest and most effective course of action?

(A) Dry the skin with paper towels.
(B) Flush the area with water and a dilute solution of a weak acid.
(C) Flush the area with a dilute NaOH solution, then water.
(D) Flush the area with water and then a dilute solution of $NaHCO_3$.
(E) Sprinkle the area with powdered Na_2SO_4, then carefully shake off the excess and rinse with water.

432. A student dissolves a 20.0 mol sample of acetic acid, a weak acid, into 1 kg of water to make a 20.0 molal solution. Assuming no other information is available to the student, which of the following is the best way to determine the *molarity* of the solution?

(A) Titrate with a standard acid
(B) Measure the total volume of the solution
(C) Determine the freezing point of the solution
(D) Measure the pH with a calibrated pH meter
(E) Measure the electrical conductivity of the solution

433. A 98-gram sample of phosphoric acid, H_3PO_4 (a weak acid), is dissolved in 1 kg of water. Assuming no other information is available, which of the following procedures is the best way to determine the *molarity* of the solution?

(A) Measure the total mass of the solution
(B) Measure the total volume of the solution
(C) Measure the pH of the solution with a calibrated pH meter
(D) Determine the boiling point of the solution
(E) Titrate the solution with a standard acid

Questions 434–438 refer to the following answer choices:

(A) Visible-light spectrophotometry (colorimetry)
(B) Paper chromatography
(C) Titration
(D) Gravimetric determination or differential precipitation
(E) Electrodes

434. Can be used to measure the conductivity of a solution

435. Used to separate pigments in a mixture

436. Used to determine the concentration of a solution of $KMnO_4$

437. Used to determine if Na^+, Mg^{2+}, and/or Pb^{2+} ions are present in a solution

438. Used to determine the unknown concentration of a known reactant

439. Which of the following procedures would allow a student to boil water at a temperature significantly above 100°C in a laboratory in which the temperature is 22°C and the pressure is 1 atm?

(A) Add salt to the water
(B) Stir or agitate the water as it is heating
(C) Insulate and cover the container in which the water is heated
(D) Heat the water in a sealed container into which air has been pumped to increase the pressure in the container
(E) Heat the water in a sealed container from which all the atmospheric air has been removed

440. Which of the following measures of concentration changes with temperature?

(A) Mole fraction
(B) Mass fraction
(C) Mass percentage
(D) Molality
(E) Molarity

441. Which of the following pieces of laboratory glassware would be used to most accurately transfer a 10.00-mL sample of solution?

(A) 5-mL pipet
(B) 10-mL pipet
(C) 10-mL graduated cylinder
(D) 15-mL graduated cylinder
(E) 15-mL Erlenmeyer flask

442. A student determined the density of a pure liquid at 25°C by measuring its mass with an electronic balance and measuring its volume in a clean, dry 50.00-mL volumetric flask. On the basis of this information and the measurements shown in the table below, to how many significant digits should this density be reported?

	Mass (g)
Empty flask	63.9892
Flask + liquid	104.4932

(A) 2
(B) 3
(C) 4
(D) 5
(E) 6

443. A student weighs out 0.20 mol of glucose to prepare a 2-M glucose solution. Which of the following pieces of laboratory equipment is most appropriate for preparing this solution?

(A) 100-mL graduated cylinder
(B) 1-L graduated cylinder
(C) 1-L Erlenmyer flask
(D) 100-mL volumetric flask
(E) 1-L volumetric flask

444. Which of the following techniques is most appropriate for recovering solid NH_4NO_3 from an aqueous solution of NH_4NO_3?

(A) Thin layer chromatography
(B) Evaporation
(C) Distillation
(D) Filtration
(E) Titration

Data Interpretation

Question 445 refers to the following situation.

A small sample of astatine was purified from an astatine containing ore in the lab. The pure astatine was stored at 25°C overnight. The next day, the sample was reanalyzed and found to have a significant quantity of bismuth. The sample was shipped to another lab for independent analysis, where the sample, six days after shipping, was found to have almost no astatine, a large quantity of bismuth, and a small quantity of lead.

445. Which of the following best accounts for the differences in analyses of the sample?

(A) The sample was not properly purified.
(B) The original sample was contaminated and some astatine was lost with each analysis.
(C) The astatine transmuted to bismuth, which transmuted to lead over the time course of analyses.
(D) The astatine sample was contaminated with bismuth during the first analysis and then the sample was further contaminated with bismuth and lead during the second analysis.
(E) The pure astatine sample was contaminated during the first analysis, but the analyses by the independent lab were mixed up with those of an unrelated sample from a completely different source.

Questions 446 and 447 refer to the following choices and the data table below.

A student forgot to label the flasks of 1.0-M concentrations of the following solutions:

(A) $Fe^{3+}{}_{(aq)}$
(B) Silver nitrate
(C) Barium chloride
(D) Mercury (I) nitrate
(E) Copper (II) nitrate

To identify each compound, the student mixed equal volumes of the 1.0-M solutions above, listed A–E, with a concentrated solution of $NH_{3(aq)}$. Each of the compounds was numbered 1–5.

Compound	Observation
(1)	A tan precipitate forms
(2)	No reaction
(3)	A pale blue precipitate in a dark blue solution
(4)	White precipitate
(5)	Orange precipitate

446. Compound 3

447. Compound 5

448. A student adds aqueous NH_3 to a solution of Ni^{2+} ions and a precipitate forms. When the student adds excess NH_3, the precipitate dissolves and produces a deep blue solution. Which of the following best explains why the precipitate dissolved in excess NH_3?

(A) NH_3 is a strong base at high concentrations.
(B) NH_3 only acts as a base at low concentrations.
(C) Ni^{2+} forms a soluble, complex ion with NH_3.
(D) Ni^{2+} solubility increases with increased pH.
(E) Ni^{2+} gets reduced by the excess NH_3, forming soluble Ni atoms.

449. In the laboratory, which of the following can produce a gas when added to 1 M HCl?

I. $Zn_{(s)}$
II. $NaHCO_{3(s)}$
III. 1 M $NH_{3(aq)}$

(A) I only
(B) III only
(C) I and II only
(D) I and III only
(E) I, II, and III

450. A solid, white crystalline substance is added to water to produce a basic solution. When a strong acid is added to the solution, a gas is liberated. Based on this information, the solid could be:

(A) NaCl
(B) NaOH
(C) $NaNO_3$
(D) Na_2CO_3
(E) Na_2SO_4

Questions 451–453 refer to the following experiment.

An experiment analyzed the ratios in which iron and oxygen combine to form different compounds. The following data were obtained.

Fe	O_2
28 g	8 g
28 g	4 g
28 g	10.5 g

451. This experiment best demonstrates which of the following chemical principles?

(A) Conservation of mass
(B) Conservation of energy
(C) The law of definite proportions
(D) The law of multiple proportions
(E) The law of stoichiometry

452. If no other information was available, which of the following could be determined from the data collected?

(A) Empirical formula
(B) Equilibrium constant
(C) Reaction order
(D) Enthalpy of formation
(E) Density

453. The compounds that were formed during the experiment were

I. FeO
II. FeO_2
III. Fe_2O
IV. Fe_2O_3
V. Fe_3O_4

(A) I, II, and IV
(B) I, III, and IV
(C) I, III, and V
(D) II, III, and V
(E) III, IV, and V

Questions 454–456 refer to the following experiment.

A student weighed 13 grams of blue $CoCl_2$ and placed it on a watch glass at room temperature. Within a few minutes, she observed that the compound turned purple. When she re-weighed the sample, it weighed 17 grams. Several minutes later, it turned red. When she weighed the sample a third time, it weighed 24 grams.

454. The $CoCl_2$ compound can be described as all of the following *except*:

(A) Deliquescent
(B) Hygroscopic
(C) Desiccant
(D) Efflorescent
(E) Hydrophilic

455. The correct formulas of the purple and red hydrates the student observed are:

(A) $CoCl_2 \cdot 1\ H_2O$ and $CoCl_2 \cdot 3\ H_2O$
(B) $CoCl_2 \cdot 2\ H_2O$ and $CoCl_2 \cdot 6\ H_2O$
(C) $CoCl_2 \cdot 4\ H_2O$ and $CoCl_2 \cdot 7\ H_2O$
(D) $CoCl_2 \cdot 4\ H_2O$ and $CoCl_2 \cdot 11\ H_2O$
(E) $CoCl_2 \cdot 17\ H_2O$ and $CoCl_2 \cdot 24\ H_2O$

456. Ionic compounds that undergo a significant color change when hydrated have which of the following properties?

(A) They contain a metal and a nonmetal.
(B) They contain a transition metal.
(C) They form only one kind of hydrate.
(D) They are more efflorescent than deliquescent.
(E) They contain polyatomic ions.

Questions 457 and 458 refer to the following experiment.

A student performed an experiment at 760 torr. He obtained a sample of an unknown, volatile liquid. He placed it in a 2.00-L Erlenmeyer flask and covered it with a lid containing a tiny pinhole. The student placed the flask in boiling water of 100°C until all the air in the flask escaped through the pinhole and all of the liquid was vaporized. He then immersed the flask in cold water to condense the gas. He dried the flask and determined the mass of the condensed vapor.

457. The mass of the condensed vapor was 3.0 grams. What is the molar mass of the liquid?

(A) 30 g mol^{-1}
(B) 46 g mol^{-1}
(C) 62 g mol^{-1}
(D) 128 g mol^{-1}
(E) 190 g mol^{-1}

458. All of the following could result in the stated deviation from the actual molar mass *except*:

(A) The student left some of the unknown substance in the liquid phase before immersing in the cold water, increasing the observed molar mass.
(B) The container's lid was not securely attached to the flask, decreasing the observed molar mass.
(C) The substance had not actually heated up to 100°C in the water bath as the student had thought, decreasing the observed molar mass.
(D) The student did not properly dry the flask before adding the unknown, increasing the observed molar mass.
(E) The student allowed some of the vapor to escape, decreasing the observed molar mass.

459. A student added 10 grams of an unknown, nonvolatile solute to 50 grams of water. At 760 torr, the solution boiled at 102°C. If the solute was known to be an ionic compound with a van't Hoff dissociation factor of 2, the molar mass of the solute is closest to:

(A) 72 g mol^{-1}
(B) 101 g mol^{-1}
(C) 204 g mol^{-1}
(D) 372 g mol^{-1}
(E) 500 g mol^{-1}

Questions 460–464 refer to the following graph.

The data below were obtained in the determination of the freezing point of a solution of naphthalene in para-dichlorobenzene. (Assume the compounds do not dissociate in the solution.)

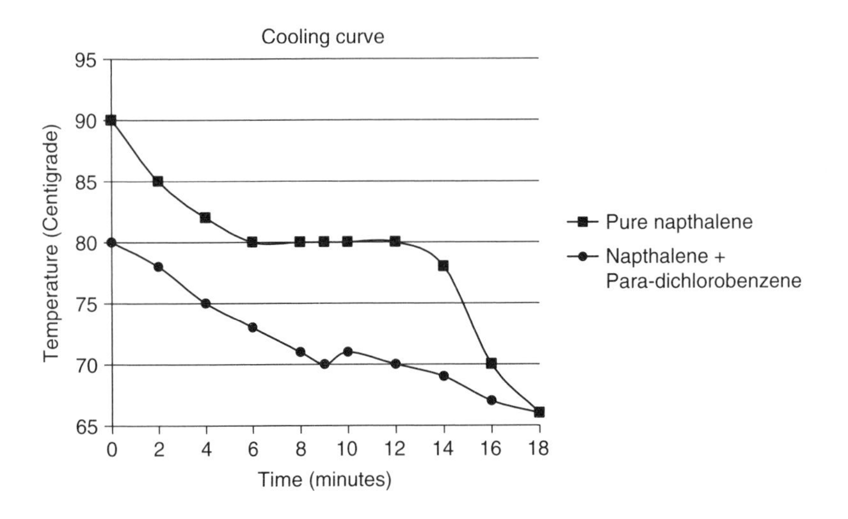

460. What is the freezing point of pure naphthalene (°C)?

(A) 90
(B) 85
(C) 80
(D) 70
(E) 65

461. The freezing point (°C) of the naphthalene-para-dichlorobenzene solution is closest to:

(A) 80
(B) 75
(C) 70
(D) 65
(E) <65

462. The molality of naphthalene is closest to which of the following? (The freezing point depression constant for para-dichlorobenzene is 7.1°C m^{-1}.)

(A) 0.71
(B) 1.4
(C) 2.8
(D) 3.6
(E) 7.1

463. A solution is prepared by dissolving 20.0 g of a nondissociating solute in 100 g of a pure solvent with a K_f = 5.0°C m^{-1} and a freezing point of –25°C. The freezing point of the solution is –37.5°C. The molar mass of the solute is closest to:

(A) 63 g mol^{-1}
(B) 80 g mol^{-1}
(C) 108 g mol^{-1}
(D) 120 g mol^{-1}
(E) 180 g mol^{-1}

464. Electromagenetic radiation in the form of X-rays can be passed through a crystal of a pure substance producing a diffraction pattern that can be captured on photographic paper and analyzed to determine molecular structure. Radiation of X-ray wavelength is used in this procedure because

(A) X-ray wavelengths are small enough to pass through the crystal.
(B) X-rays wavelengths are significantly greater than the size of the atoms so they can easily be reflected off the surface.
(C) diffraction patterns emerge from crystals only when the wavelength of the radiation is comparable in size to the distance between the atoms.
(D) the energy of X-rays is high enough to break the crystal apart and scatter the atoms into a pattern on the photographic paper.
(E) the energy of X-rays is large enough to ionize the crystal but not to break apart the bonds between atoms.

465. A student determined the percentage of water in a hydrate to be 26 percent by weighing the sample, heating it until dry, and then reweighing it. The accepted value for the percentage of water in the hydrate is 45 percent. Which of the following is the best explanation for the difference between measured and accepted values?

(A) The student had a high-percent error.
(B) The student started with a mass of sample that was too low to accurately weigh.
(C) Overheating of the sample caused some of the solid to spatter out of the crucible.
(D) Overheating of the sample caused the dry sample to decompose into a gas.
(E) The dehydrated sample absorbed moisture from the atmosphere between drying and reweighing.

466. In a laboratory, H_2 gas can be produced by adding which of the following to a 1-M HCl solution?

I. $Mg_{(s)}$
II. $Zn(NO_3)_{2(s)}$
III. $Na_2CO_{3(s)}$

(A) I only
(B) II only
(C) III only
(D) I and II only
(E) II and III only

Questions 467 and 468 refer to the answer choice listed in the table below and the following titration curve.

	Indicator	pH Range of Color Change
(A)	Phenolphthalein	8.2–10.0
(B)	Bromthymol blue	6.0–7.6
(C)	Methyl red	4.4–6.2
(D)	Methyl orange	3.1–4.4
(E)	Thymol blue	1.2–2.8

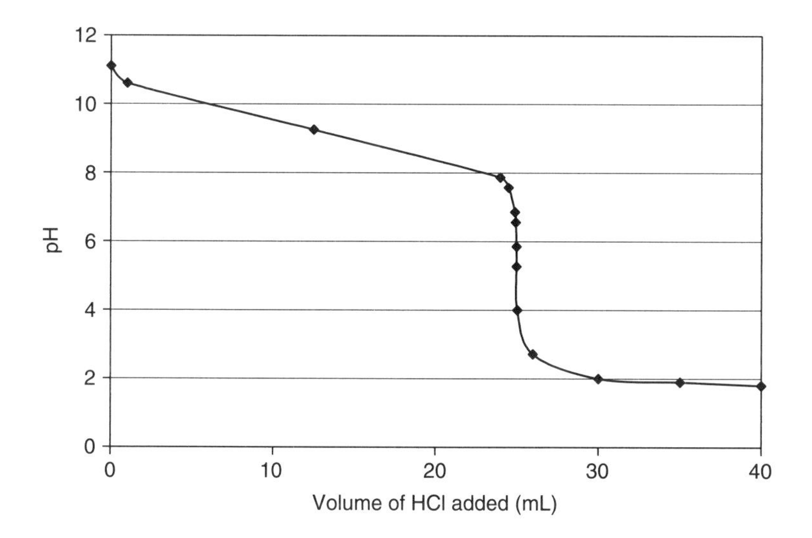

467. Which of the above indicators is the most appropriate choice for this titration?

468. Which of the above indictors would transition in the buffer region of the curve?

Question 469 refers to the following analysis.

A colorless solution was aliquoted into 3 test tubes. The following tests were performed:

Sample	Test	Observation
1	Add $HCl_{(aq)}$	No change
2	Add $NH_{3(aq)}$	No change
3	Add $SO_4^{2-}{}_{(aq)}$	No change

469. Which of the following ions could be present in the solution at a 0.15-M concentration?

(A) CO_3^{2-}
(B) Na^+
(C) Ca^{2+}
(D) Ni^{2+}
(E) Ba^{2+}

Questions 470–472 to refer to the following experimental procedure.

In a laboratory experiment, H_2 gas is produced by the following reaction:

$$Zn_{(s)} + 2\ HCl_{(aq)} \rightarrow ZnCl_{2(aq)} + H_{2(g)}$$

The H_2 gas is collected over water in a gas-collection tube (eudiometer). The atmospheric pressure in the laboratory is 770 torr and the temperature of the lab and the water used in the experiment is 22°C. The vapor pressure of water is 19.8 torr at 22°C.

470. Before measuring the volume of gas collected in the tube, which of the following is a necessary step in correctly determining the total gas pressure inside the tube?

(A) Increase the temperature of the water to 25°C.
(B) Wait for the atmospheric pressure in the lab to reach 760 torr.
(C) Let air into the tube to break the vacuum.
(D) Adjust the tube so that the water level inside the tube is the same as the water level outside the tube.
(E) Lift the tube until the lip is just barely immersed in the water.

471. Which of the following gases should not be collected using this technique?

I. HCl
II. NH_3
III. CO_2

(A) I only
(B) II only
(C) III only
(D) I and II only
(E) II and III only

472. The partial pressure of hydrogen gas in the tube is closest to:

(A) 730 torr
(B) 750 torr
(C) 760 torr
(D) 770 torr
(E) The partial pressure cannot be determined without knowing the volume of gas collected

Questions 473–475 refer to the following experiment.

A student designed a procedure to determine the heat of fusion of ice. She constructed a calorimeter using a polystyrene cup and a thermometer. She weighed the cup, filled it with 150 mL of warm water, then weighed the cup and water together. The temperature of the water was measured, and then ice taken from an ice bath temporarily stored in a –20°C freezer was added to the cup. The cup with water and ice was weighed. The student then covered the cup with a polystyrene lid with two small openings. In one opening, she inserted a thermometer and in the other, she inserted a stirring rod to gently stir the contents of the cup until all the ice was melted. The lowest temperature reached by the water in the cup was recorded.

473. The purpose of weighing the cup and its contents the third (last) time was to

(A) determine the mass of water that was added.
(B) determine the mass of ice that was added.
(C) determine the mass of ice and water that were added.
(D) determine the mass of the calorimeter and thermometer.
(E) determine the mass of water that evaporated while the lid was off the cup.

474. Suppose a significant amount of water from the ice bath adhered to the ice cubes that the student added to the calorimeter. How would this affect the value of the heat of fusion of ice is calculated?

(A) There would be no effect because the water from the bath would be at the same temperature as the ice cubes.
(B) The calculated value would be too large because less warm water needed to be cooled.
(C) The calculated value would be too large because more cold water needed to be heated.
(D) The calculated value would be too small because more cold water needed to be heated.
(E) The calculated value would be too small because less ice melted than was weighed.

The student's final data is recorded below.

$T_{initial}$ of water	25°C
Mass of cup	6.40 g
Mass of cup + water	156.50 g
Mass of cup + water + ice	181.56 g
T_{final} of water	14°C

475. Which of the following pieces of information is necessary to calculate the heat of fusion of ice from this data?

I. The specific heat of water
II. The specific heat of ice
III. The thermal conductivity of water
IV. The thermal expansion coefficient of ice

(A) I only
(B) II only
(C) I and II only
(D) I and III only
(E) I, II, and IV only

476. What mass of KOH (molar mass 56 g mol^{-1}) is required to make 250 mL of a 0.400 M KOH solution?

(A) 1.00 g
(B) 5.60 g
(C) 8.96 g
(D) 14.0 g
(E) 22.4 g

477. A steady electric current is passed through molten NaCl for exactly 2 hours producing 230 g of Na metal. The same current is passed through molten $FeCl_3$ for exactly 2 hours. The mass of Fe metal expected to be produced is closest to:

(A) 56 g
(B) 112 g
(C) 168 g
(D) 186 g
(E) 224 g

478. When 100 mL of 2 M $Pb(NO_3)_2$ is mixed with 100 mL of 3 M NaCl in a beaker, a white precipitate forms. Which of the following is true of the concentration of ions remaining in solution?

(A) $[Na^+] > [NO_3^-] > [Pb^{2+}] > [Cl^-]$
(B) $[NO_3^-] > [Na^+] > [Cl^-] > [Pb^{2+}]$
(C) $[Na^+] > [NO_3^-] > [Cl^-] > [Pb^{2+}]$
(D) $[NO_3^-] > [Na^+] > [Pb^{2+}] > [Cl^-]$
(E) $[Na^+] > [Cl^-] > [NO_3^-] > [Pb^{2+}]$

479. Which of the following would be a qualitative test for a solution of Ba^{2+}, Fe^{3+}, and Zn^{2+} ions that would separate Ba^{2+} from the other ions at room temperature?

I. Adding dilute HCl
II. Adding dilute NaOH
III. Adding dilute Li_2SO_4

(A) I only
(B) II only
(C) III only
(D) II and III only
(E) I, II, and III

A discharge tube filled with only hydrogen gas was electrified. The gas gave off blue light, which was polarized and then passed through a prism. Four narrow, colored bands were observed on a screen behind the prism. The energy of a photon is given by the equation $E = h\nu$, where $h = 6.63 \times 10^{-34}$ J·s and ν = the frequency.

The following data were collected during the experiment.

Band	Color	Wavelength, λ (10^{-9}m)	Frequency, ν (sec^{-1})
1	Violet	410	7.3×10^{14}
2	Blue violet	434	6.9×10^{14}
3	Blue green	486	6.2×10^{14}
4	Red	656	4.6×10^{14}

480. Which of the following best explains why hydrogen gas emitted light when electrified?

(A) For energy to be conserved in an atom, photons are emitted when an electron drops to ground state after being excited.
(B) Electrons absorbed photons of electricity that provided the energy needed for them to be ejected.
(C) The ionized gases produced by the electric current emit photons.
(D) The electricity caused the gas particles to collide with great kinetic energy, producing photons.
(E) The electrons turned into photons when subjected to an electric field.

481. Wave-like properties of light include which of the following?

I. Interference
II. Polarization
III. The photoelectric effect

(A) I only
(B) II only
(C) III only
(D) I and II only
(E) II and III only

482. In another famous experiment, a metal plate was bombarded with photons of different frequencies. At frequencies above 4.4×10^{14} sec^{-1}, electrons were ejected from the metal, hence ionizing it. Which of the following is closest to the ionization energy of the metal in Joules?

(A) 1.5×10^{-48}
(B) 6.6×10^{-34}
(C) 4.4×10^{-20}
(D) 2.9×10^{-19}
(E) 3.0×10^{-14}

483. A photon of red light is produced by an atom. Which of the following expressions accurately calculates its energy?

(A) (4.6×10^{14}) (656)
(B) (4.6×10^{14}) (6.63×10^{-34})
(C) (4.6×10^{14}) (3×10^{8})
(D) (656) (6.63×10^{-34})
(E) (656) (6.63×10^{-34})

484. All of the following are true statements regarding atomic spectra *except:*

(A) Line spectra are typical of electrified gases and continuous spectra are produced from the glow of hot objects.
(B) The electron configuration of the atom determines the type of spectra that will be emitted.
(C) The number of lines in the spectra is directly proportional to the number of electrons in the atom.
(D) Photons with lower wavelengths than those of visible light can be can be emitted by atoms.
(E) The lines produced in atomic spectra support the quantum mechanical model of that atom that says there are achievable energy states.

485. Which of the following is true regarding the polarization of the blue light in the experiment?

(A) Monochromatic light (one color) must be polarized in order to pass it through a prism.
(B) The polarizer focuses the light so it hits the prism in a more intense band.
(C) Polarization filters waves of light so that only waves oriented in the same plane emerge from the polarizer.
(D) The polarizer is a weak prism that partially separates the different wavelengths of light so that the prism can separate them more effectively.
(E) The polarizer absorbs wavelengths of light that are not in the visible range to prevent them from entering the prism.

Questions 486–488 refer to the following experiment.

A beam of gaseous hydrogen atoms is emitted from a hot furnace and passed through a magnetic field onto a detector screen. The interaction of the electron of the hydrogen atom and the magnetic field causes the hydrogen atom to be deflected from a straight line path. A tiny, permanent spot develops where an atom strikes the screen.

486. If the spin of the electron in a hydrogen atom (only 1 electron) was completely random, which of the following patterns would be observed on the screen?

(A) One very small, focused spot in the middle of the screen
(B) A very dark spot in the center of the screen that becomes more diffused as the radius increases
(C) No discrete spots would be observed
(D) Two spots of equal intensity
(E) A line in the shape of a wave

487. Which of the following observations was the experiment designed to further study?

(A) The energy of atoms is quantized.
(B) Electrons have both wave- and particle-like properties.
(C) It is impossible to know both the position and momentum of an electron with certainty.
(D) Electrons have a charge equal in magnitude but opposite in charge to protons, but their mass is 1/1,800 that of a proton.
(E) Emission spectra lines of hydrogen and sodium can be split when a magnetic field is applied.

488. The data produced in this experiment verify which of the following concepts of the atom?

(A) Electron energies are quantized.
(B) Electron spin is either $+\frac{1}{2}$ or $-\frac{1}{2}$.
(C) Electrons occupy orbits around the nucleus.
(D) Electrons have properties of both waves and particles.
(E) The more that is known about an electron's position, the less can be known of its momentum (or velocity).

489. Particle-like properties of light include which of the following?

I. Interference
II. Polarization
III. The photoelectric effect

(A) I only
(B) II only
(C) III only
(D) I and II only
(E) II and III only

Questions 490–493 refer to the following experiment.

A cathode ray strikes a detector in a straight line, but when a magnetic or electric field is applied, the path of the ray is deflected.

490. Accurate interpretations of this observation include:

I. Cathode ray particles are charged.
II. Cathode rays have both wave- and particle-like properties.
III. Cathode rays are composed of electrons.

(A) I only
(B) II only
(C) III only
(D) I and II only
(E) I and III only

491. The experiment revealed a charge-to-mass ratio of -1.76×10^8 C g^{-1}. The charge of an individual electron is -1.602×10^{-19} C. Which of the following correctly expresses the mass of an individual electron?

(A) $(-1.602 \times 10^{-19})(-1.76 \times 10^8)$
(B) $(-1.602 \times 10^{-19})(-1.76 \times 10^8)^{-1}$
(C) $(-1.76 \times 10^8)(-1.602 \times 10^{-19})^{-1}$
(D) $(-1.602 \times 10^{-19})(-1.76 \times 10^8)^{-1}(6.02 \times 10^{23})$
(E) $(-1.602 \times 10^{-19})(-1.76 \times 10^8)(6.02 \times 10^{23})$

492. Which of the following particles would *not* be deflected when passed through an electric field?

(A) α (alpha) particle
(B) β (beta) particle
(C) Proton
(D) Neutron
(E) Positron

Questions 493–498 refer to the following data collected at 20°C.

	Liquid	Viscosity (cP)	Surface Tension ($^{N}/_{m}$)
(A)	Benzene	0.652	0.0289
(B)	Water	1.0	0.0728
(C)	Olive oil	84	0.0320
(D)	Castor oil	986	?
(E)	Glycerol	1,490	0.0634

493. The liquid with the steepest meniscus in a glass tube of small diameter

494. The liquid with the greatest resistance to flow

495. The liquid with the greatest intermolecular forces of attraction

496. Which pair of liquids listed below is miscible?

(A) Benzene and water
(B) Water and olive oil
(C) Castor oil and glycerol
(D) Benzene and glycerol
(E) Water and glycerol

497. The expected surface tension (in $^{N}/_{m}$) of castor oil is approximately:

(A) <0.02
(B) Between 0.020 and 0.032
(C) Between 0.033 and 0.064
(D) Between 0.065 and 0.073
(E) >0.074

498. Viscosity and surface tension decrease with increasing temperature. This is most accurately explained by the fact that

(A) liquids expand when heated, increasing the distance between the molecules.
(B) both viscosity and surface tension are primarily determined by the intermolecular forces of attraction between the molecules in the liquid.
(C) density decreases with increasing temperature, allowing the molecules to flow more freely past each other.
(D) increasing temperature increases reaction rates.
(E) increasing temperature increases the K_{eq} of the equilibrium state of both the viscosity and surface tension.

Questions 499 and 500 refer to the following answer choices:

Compound	Molarity of Compound in Water (mol L^{-1})	Molarity of Particles in Water (mol L^{-1})
(A)	1.000	1.000
(B)	1.000	1.006
(C)	1.000	1.500
(D)	1.000	2.000
(E)	1.000	3.000

499. A weak acid

500. The best electrolyte

ANSWERS

Chapter 1: Atomic Theory and Structure

1. (B) The *mass number* of cadmium is 112, not the atomic mass (the weighted average of the naturally occurring isotopes). The mass number will always be a whole number because it is the sum of the number of protons and neutrons (collectively called the *nucleons*, referring to their location in the nucleus) in an atom. The number of electrons and protons will always be the same in a neutral atom because they are the only negatively and positively charged (respectively) particles in the atom. *The atomic number is the number of protons.* It determines the identity of the atom, so finding cadmium on the periodic table will tell us its atomic number. If we subtract the atomic number from the mass number, we get the number of neutrons in that particular isotope.

2. (E) Location E on the periodic table is in the vicinity of francium. Francium has a half-life of just 22 minutes and is the second rarest naturally occurring element (astatine is the rarest). It is doubtful anyone has actually reacted francium with water, but if it behaves as expected, it would certainly be a spectacle. In general, but moreso for the alkali metals at the bottom of the group (with the largest atomic radii and the lowest first ionization energies), the reaction is highly exothermic, partly because the reaction produces a strong base whose total dissociation in water is highly exothermic. The reaction also produces energy in the form of light and $H_{2(g)}$.When added to water, the interior of a piece of sodium metal, for example, will melt before it is consumed due to the high temperature produced by the reaction (the melting point of Na is ~800°C). In addition, the high heat ignites the flammable $H_{2(gg)}$ that is produced.

3. (B) The location of B on the table is in the halogens, specifically fluorine. Fluorine does not have the highest ionization energy of *all* the elements, just the elements we've been given to choose from in this question. *Helium is the element with the highest first ionization energy. First ionization energy* is the minimum amount of energy required to ionize a ground state, gaseous atom by removing an electron. The result is a cation. In general, *the first ionization energy is low for metals and high for nonmetals.* (See the figure in **Questions 21 and 22** to compare the first ionization energies of elements 1–20.)

4. (B) The location of B on the table is in the vicinity of the halogens, specifically fluorine. *Fluorine is the element with the highest electronegativity. Electronegativity is a measure of an atom's ability to attract electrons to itself while in a bond (within a molecule).* Because the measurement of electronegativity of an atom relies on the atom being in a bond, the noble gases He, Ne, and Ar do not have measured values for electronegativity. The ionization energies of the valence electrons in Kr, Xe, and Rn are sufficiently low, allowing these noble gases to form covalent bonds with other atoms (mostly those with a high electronegativity, like F, whose electron-attracting abilities are strong enough to force the Kr, Xe, or Rn atoms to share their electrons) and therefore have their electronegativities assessed.

5. (C) *Electron affinity is the measure of the energy change that occurs when an electron is added to a ground state, gaseous atom.* Atoms that have a high affinity, or attraction, for electrons have very negative electron affinities. The periodic trend for electron affinity is generally correlated with electronegativity, so regions B and C are the top contenders for answer choices. However, there are some important differences. All the noble gases have measured electron affinity (the lightest three have no values for electronegativity, see **Answer 4**) and *chlorine, not fluorine, has the highest electron affinity.* Electron affinity doesn't change significantly within a group, and it can be thought of as the reverse ionization energy of an atom's −1 anion. For example, the energy change to remove an electron from Cl^- (to produce Cl) is 349 kJ mol^{-1} (endothermic), and the energy change to add an electron to Cl to produce Cl^- is −349 kJ mol^{-1} (exothermic).

6. (E) Location E on the periodic table is in the vicinity of francium, which might have the largest atomic radius if it were known (with a half-life of just 22 minutes, it doesn't exist long enough to do the necessary measurements). Atoms don't have a sharp, well-defined edge, so their *bonding atomic radius* is used to infer the radius of an individual atom. The atomic radius is measured while the atom is in a bond with another atom of the same kind. The distance between the two nuclei is measured and then divided in half. What we do know is that *cesium has the largest atomic radius of the elements measured so far.* (See **Answer 10** for an explanation of the periodic trend regarding atomic radius.)

7. (E) Metallic character is not something that is specifically measured. It is a set of properties given to metals, but the properties are due to one of the *most basic properties of metals—their readiness to lose electrons.* This is due to *metallic bonding,* which can be thought of as the most "sharing" form of bonding. *Metals are lattices of positively charged ions that are bathing in a sea of electrons.* These *electrons are highly mobile* and account for nearly all the properties of metals, especially their electrical conductivity. The weak pull on the valence electrons by the nucleus allows them to be pulled off easily, resulting in the *low ionization energies* and the relatively strong tendency of metals to lose electrons and take on (almost) exclusively positive oxidation states. Metals with the highest metallic character can be considered as those having the lowest ionization energies, though this is a bit of a simplification (but it will work for the AP Chemistry exam).

8. (C) The atomic mass of bromine is almost 80. The two isotopes are of masses 79 and 81. The average of these two numbers is 80, and that implies the masses were equally weighted in the calculation. Remember, different (naturally occurring) isotopes of atoms exist in different quantities. These are accounted for in the atomic mass calculation according to their percent natural occurrence. Remember, the atomic mass is the weighted average of *all* the naturally occurring isotopes of an element. (**Question 9** is similar.)

9. (C) The three isotopes of Sr are 86, 87, and 88. If they occurred in equal numbers, their atomic mass would be the average of their mass numbers, 87. Since the mass number is *greater* than the average, the isotopes of higher mass must be present in greater quantities. Additionally, because the atomic mass is closer to 88 than 87, we can predict the occurrence of the 88 isotope as the highest. Although (E) is mostly true—the natural occurrence of isotopes 86 and 87 are 9.9 percent and 7.0 percent, respectively, and that cannot be established from the information given in the question. (**Question 8** is similar.)

10. (D) Atomic radius decreases from left to right along a period. This is due to the shielding effect of the core electrons (this doesn't apply to atoms with one electron, such as hydrogen, or a He^+ ion). All the elements in a particular period have the same configuration of core electrons. These electrons shield the valence electrons from the pull of the nucleus. The effective nuclear charge is calculated $Z_{eff} = Z - S$ (Z_{eff} is the effective nuclear charge; Z is the atomic number, a.k.a. the number of positive charges in the nucleus; and S is the number of nonvalence, or core, electrons). The Z_{eff} for all the atoms in a period gets larger as the number of positive charges in the nucleus increases, but the number of nonvalence (core) electrons doesn't. An increased effective nuclear charge means the valence electrons feel a greater pull from the nucleus, and can thus be found closer to the nucleus than the electrons in atoms with lower values of Z_{eff}. (See **Answer 6** for an explanation of how atomic radius is measured.)

11. (C) The alkali metals form strong bases when they react with water, not strong acids. (See **Answer 2** for a description of the reaction of alkali metals with water.)

A general strategy for *except* questions: The *except* questions are tricky, even if what they are asking is not. Our brain doesn't think in the negative, so a good habit to get into is to circle the word *except* in the question to remind us that we are looking for a false statement, then treat each answer choice as either true or false, marking each choice as we go. At the end of choice (E), we choose the false one as our answer.

12. (D) This question is asking if we know that ground state elements in the same group have similar properties. Phosphorus and astatine are both in group 15 (5A) so their valence shell electron configurations are both s^2p^3, conferring on them similar chemical reactivities. Sulfur, selenium, and oxygen are in group 16 (s^2p^4), while silicon is a group 14 semimetal (metalloid) with valence shell electron configuration of s^2p^2.

13. (B) This has a simple mathematical solution—take the atomic number (which will tell us the number of electrons in a neutral atom) of the element and *add* the absolute value of the *negative* oxidation states (more electrons) and *subtract* the absolute value of the *positive* oxidation states. F^- (9+1) and Na^+ (11−1) both have 10 electrons and are therefore isoelectronic.

We can also arrive at this answer by finding one of the elements in each pair on the periodic table and moving one element to the right for each negative charge and one to the left for each positive charge. If the two elements we are comparing lead us to the same element once we've accounted for their oxidation state, then they've got the same number of electrons. For Na^+ and F^-, this element would be neon. The same number and configuration of electrons does *not* correlate with similar chemical reactivity in ions. F^- and Na^+ are like neon in that they have a full valence shell and both are more stable and less reactive than in their ground state, but they are charged and therefore behave like ions. Their ionic radius also differs due to their different nuclear charge (see second paragraph of **Answer 15**).

14. (D) See **Answer 13**, keeping in mind that the iodide ion has 54 electrons.

15. (C) Ionic radius is *not* the same as atomic radius (described in **Answer 6**). In an ion, the number of electrons *does not* equal the number of protons. Atoms become ions because they gain or lose electrons (not protons), so ions that are positively charged will be smaller than what they are when in their ground state (same effective nuclear charge pulling on fewer electrons), whereas negatively charged ions will be larger when they are in their ground state (same effective nuclear charge pulling on more electrons).

For isoelectronic ions (ions with the same number of electrons), the ionic radius decreases with increasing nuclear charge. For example, $O^{2-} > F^{-} > Na^{+} > Mg^{2+} > Al^{3+}$. This is because the same number of electrons are being pulled by an increasing number of protons. (See **Answer 10** for an explanation of how size is affected by shielding.) For atoms of the same charge (and in the same group), the size of ions increases as you go down the group. Keep in mind that ionic size is an important determinant of lattice energy (see **Answer 52** for a description of the factors that affect lattice energy).

16. (D) The first ionization energies of Kr, Xe, and Rn are sufficiently low to allow these noble gases to form covalent bonds with other atoms. Krypton difluoride, KrF_2, was the first compound of krypton discovered. Xenon can form compounds with oxygen (XeO_3 and XeO_4) and fluorine (XeF_4 and XeF_6). Radon appears to form compounds with fluorine (RnF_2). Notice that oxygen and fluorine are highly electronegative atoms. It is their strong electron attracting abilities that force Kr, Xe, and Rn atoms to share their electrons and form covalent bonds. (See **Answer 11** for an *except* question strategy.)

17. (B) The operative word in this question is *diatomic.* The noble gases are monatomic and so any answer choice that contains a group 18 gas is incorrect. With the exception of astatine, which is more metallic than the rest of the elements in the group, all of the halogens (group 17) are diatomic in their standard states, but only F_2 and Cl_2 are gases. Br_2 is a liquid and I_2 is a solid.

18. (C) The actual pattern of atomic size is not as regular as our general trend describing it. The transition elements present some exceptions. For example, the atomic radius of the manganese group (7) and the copper group (11) have larger atomic radii than those to the right and left of those elements. But for elements with *only* s and p outer elections, the atomic size decreases only from left to right.

19. (D) Ionization energy is an indicator of effective nuclear charge. (See **Answer 10** for an explanation of effective nuclear charge and the figure accompanying **Questions 21 and 22** for a graph of ionization energies.) The other choices are incorrect because (A) has nothing to do with effective nuclear charge, (B) and (C) are false, and (E) states the *opposite* effect of shielding on ionization energy. Less shielding = higher ionization energy. (Because the effective nuclear charge is larger with less shielding, the nucleus pulls more strongly on the electrons. This is evident because more energy is required to *remove* the electron.)

20. (D) The table lists successive ionization energies, the minimum energy requirements for the further ionization of an element (by removal of successive electrons). From the table, we see that 786 kJ of energy per mol of silicon is required to remove the first electron (a p^2 electron), leaving Si^{+}. Removing *another* electron (the p^1) requires an additional 1,577 kJ for per mol Si^+. The trick to answering this kind of question is to find a very large "jump" in ionization energies. For silicon, it's between the fourth and fifth ionizations. This indicates that the fifth ionization energy is "digging into" the core electrons because all the valence electrons have been removed. Now we know we're basically looking for an element with four valence electrons. The other elements in group 14 will show a similar trend in successive ionization energies, but the absolute numbers will, of course, vary from lower than those for Si (Ge, Sn, Pb) and higher for C.

21. (D) Elements of atomic numbers 2, 10, and 18 are He, Ne, and Ar, respectively. These noble gases have the highest first ionization energies. The large drop in ionization energy is mainly because the elements of atomic number 3 (Li), 11 (Na), and 19 (K) have s^1 electrons that are far from the nucleus. It is both this large radius and the lower effective nuclear charge (which is one of the reasons for the large radius) that make the energy requirements for removal of this electron so low. Generally, the size of the atom indicates the strength of the nucleus' pull on the electrons.

We need to be careful, however, as there is another way to think about this: The force of an electric field produced by a charged particle is inversely related to the square of the distance. In other words, double the distance, and the force decreases by one-fourth. The effective nuclear charge is *not* the only determinant of an atom's size; for larger atoms, we must also consider that the distance between an electron and the nucleus will significantly affect the force of the pull experienced by the electron.

Electron affinity is a measure of the energy change that occurs when an electron is added to a ground state, gaseous atom. Atoms 2, 10, and 18 don't have a high electron affinity because they have full valence shells and much energy must be added to overcome the repulsion of the electrons already in the atom (like charges repel). Atoms that have a high affinity, or attraction, for electrons have very negative (exothermic) electron affinities. The periodic trend for electron affinity in generally correlated with electronegativity, but with some important differences: (1) All the noble gases have measured electron affinity (though the lightest three have no values for electronegativity) and chlorine, *not fluorine*, has the highest electron affinity. (2) Electron affinity doesn't change much within a group. (3) Electron affinity can be imagined as the reverse ionization energy of an atom's −1 anion. For example, the energy change to remove an electron from Cl^- (to produce Cl) is 349 kJ mol^{-1}, and the energy change to add an electron to Cl to produce Cl^- is −349 kJ mol^{-1}. (See **Answer 3** for more on first ionization energy.)

22. (D) Statements I and II are correct. Statement III is not entirely correct because filled orbitals are not necessarily more stable. For example, the drop between elements 7 and 8 is due to the repulsion of the second electron in the p_x orbital. Having one electron in each p_x orbital is more stable than having two electrons in one orbital and one in the other 2 (the p_y and p_z orbitals). The information we need to answer the question is all in the graph, even if we don't know *why*. All we need to do is compare the ionization energy with the electron configuration using our periodic table.

23. (C) See **Answers 13 and 15** for strategies to compare atomic and ionic radii.

24. (B) We need to look for electron configurations that have electrons missing from lower energy subshells. We need to be careful, however, because the people who write the AP Chemistry exam often list 3d *before* 4s, and so it looks like the 4s is the highest energy subshell if we're not paying attention. In choice (B), the 2p subshell has only five electrons instead of six, and yet the 3s subshell is filled. This indicates that a 2p electron jumped into the 3s orbital (which was already occupied by one electron) by absorbing energy.

25. (A) Gallium is a group 13 element, meaning its one valence electron is in the p_x orbital.

26. (C) Carbon has two p electrons, each of which is unpaired in their respective p_x and p_y orbitals.

27. (C) We need to make sure we don't confuse energy levels (n = 1, 2, 3, etc.) with the s, p, d, and f sublevels (subshells). Carbon has two p electrons, each of which is unpaired in their respective $2p_x$ and $2p_y$ orbitals.

28. (B) Technetium has no stable isotopes and is the atom of lowest atomic number for which that is true. Nearly all Tc is produced synthetically. Naturally occurring Tc is produced by fission in uranium or by neutron capture by molybdenum.

29. (C) Sodium, and all the group 1 alkali metals, are highly reactive in their ground state. In particular, they react with water (even in the atmosphere, which is why alkali metals are stored in oil) to form H_2 gas and a strong base (NaOH in this case). (See **Answer 2** for a more detailed description about the reaction of the alkali metals with water.)

30. (A) Helium. Because of the small radius, helium's complete and stable valence shell of electrons, and its high effective nuclear charge, He has the highest first ionization energy of all the elements.

31. (D) We are looking for an atom with an incomplete lower subshell or orbital. Atom D, sodium, has two electrons in its 3s subshell, but only one electron in the 2s subshell.

32. (B) *Boron is an exception to the octet rule.* The mnemonic "B is happy with 3" reminds us that boron can form compounds with just three bonds joining it to the other atoms. In these compounds (like BH_3), the boron atom has no lone electrons, so both the electron and molecular geometry are trigonal planar. (See **Answer 62** for more on boron.)

33. (E) Nitrogen. Only count the electrons in the s and p orbitals of the highest (but same) energy level (n) as valence electrons.

34. (E) Nitrogen. The standard state form of nitrogen, N_2, makes up about 78 percent (mol fraction) of the earth's atmosphere.

35. (A) Helium. Atoms form compounds to complete their valence shell. All the noble gases have a complete valence shell, which is why they are monatomic gases under standard conditions.

36. (A) An excited hydrogen atom.We are looking for an atom with an incomplete lower subshell or orbital. Atom A has an electron in its 2s subshell, but not the 1s orbital.

37. (E) Cobalt. The cations of transition elements produce colored compounds and solutions, so we are looking for elements in which the electrons of highest energy level (which may not be the highest in value) are in d orbitals. Cobalt produces magenta or blue solutions depending on its oxidation state (II or III, respectively).

38. (C) Neon. An unreactive atom is one in which the valence shell of the atom is filled, in other words, a noble gas.

39. (D) Sodium. Any of the group 1 metals react violently with water according to the following reaction:

$$Na_{(s)} + 2\ H_2O_{(l)} \rightarrow NaOH_{(aq)} + H_{2(g)}$$

The reaction produces a strong base whose total dissociation in water is extremely exothermic and the high temperature it produces ignites the flammable $H_{2(g)}$ that is produced. (See **Answer 2** for more on the reaction of alkali metals with water.)

40. (D) Sodium. The atom with the highest second ionization energy is the one in which removing the first electron leaves a full valence shell behind. Choice (D), sodium, has *one* 3s electron. Removing that electron leaves the neon electron configuration behind (but with a higher effective nuclear charge, so the second ionization energy of Na is *higher* than the first ionization energy of neon).

41. (E) Cobalt. (See **Answer 37.**)

42. (D) Sodium. The alkali metals are highly reactive due to their low first ionization energies.

43. (B) Rutherford bombarded a thin sheet of gold foil with alpha particles (He^{2+}, helium nuclei). Most of the He^{2+} passed straight through the foil, indicating that the atoms making up the foil were mostly empty space. Some of the He^{2+} particles were deflected from their paths, but a few actually backscattered. These deflected and backscattered He^{2+} particles suggested that the positive charges of an atom were concentrated into a small volume, hence the great repulsion when the He^{2+}approached. Bohr conceived of the energy levels of electrons. Choice (C) is not true, and several scientists contributed to the fact stated in choice (D).

44. (E) With the exception of astatine, which is more metallic than the rest of the elements in the group, all of the halogens (group 17) are diatomic in their standard states. Only F_2 and Cl_2 are gases. Br_2 is a liquid and I_2 is a solid. Astatine is the rarest of the naturally occurring elements, and is therefore not usually considered with the rest of the halogens, so we don't use astatine to make exceptions for the halogen group.

45. (C) Metals practically *never* take on negative oxidation states. Nonmetals can take on negative or positive oxidation states. Compared to the other elements, the halogens have high electron affinities and most often take on a −1 oxidation state (fluorine *always* takes on a −1 oxidation state).

46. (A) A monovalent cation is a +1 cation. The alkali metals have 1 valence electron and low first ionization energies. They exclusively form +1 cations.

47. (B) First ionization energy is the minimum amount of energy required to ionize a ground state, gaseous atom by removing an electron. The result is a cation. In general, the first ionization energy is low for metals and high for nonmetals. Helium is the element with the highest first ionization energy. (See the figure accompanying **Questions 21 and 22** to compare the first ionization energies of elements 1–20).

48. (E) All naturally occurring atoms with atomic number 84 or above (polonium and higher) are radioactive. The naturally occurring actinides have atomic numbers 89–92 (actinium, thorium, protactinium, and uranium, respectively).

49. (B) Oxidation is loss of electrons (the mnemonic OIL RIG is helpful: **O**xidation **I**s **L**oss of electrons, **R**eduction **I**s **G**ain of electrons). Generally, the first ionization energy of an atom indicates how difficult an atom is to oxidize. The noble gases have the highest first ionization energies and are the most difficult to oxidize, mostly because the valence electrons in the noble gases experience the highest effective nuclear charge in their respective periods.

50. (E) Substance 1 contains Ca^{2+} and Substance 2 contains Cu^{2+}. Although we should be familiar with the flame test, the real clues to the question are in the solutions. The alkali and alkali earth metals produce colorless solutions and transition metals produce colored solutions. It's worth memorizing that Cu^{2+} produces a blue solution (use the mnemonic "Copper 2 in water blue").

Chapter 2: Chemical Bonding

51. (D) Ionic character refers to a place on the continuum of bond character. On one end of this continuum is covalent character. This is the perfectly equal sharing of electrons between atoms. This occurs only when two of the same atoms are bonded together, F_2, for example. Covalent character begins to decrease as the sharing of the electrons becomes less equal. This can be predicted by comparing the electronegativities of the atoms involved in the bond. The greater the difference in electronegativity between the two atoms, the less equal the sharing of electrons is until the difference gets so large, that sharing is no longer an option. On the opposite side of the bond character continuum is ionic character. An atom with a very low electronegativity, like Cs, doesn't exert a strong pull on its valence electron, the one involved in the bond. Fluorine has the highest electronegativity and pulls very strongly on the electrons in the bond it is a part of. If cesium is not strongly attached to its electron, fluorine is happy to take it. So Cs loses an electron to F, becoming Cs^+ and F^-, respectively. Now, the opposite charges on these two ions will cause them to attract each other very strongly in a bond of high ionic character.

The bottom line is that the greater the difference in the electronegativity between the two atoms, the more ionic character the bond, or the compound, has. The compounds in choices (B), (C), and (E) are all covalent compounds. Ionic compounds are typically (though not always) a metal (low electronegativity) and a nonmetal (high electronegativity). Silicon is a semimetal and aluminum is a metal. The difference in electronegativity between Al and F is about 2.4, whereas the difference in electronegativity between Si and O is about 1.5. (This still indicates some ionic character. Some people draw the line for an ionic bond as an electronegativity difference of about 2, though that number varies and can be as low as 1.7.)

However, we should recognize SiO_2 (silicon dioxide, also known as quartz) as a network solid, a solid in which all of the atoms are covalently bonded in a continuous network.

52. (E) Lattice energy is the minimum energy required to completely separate the ions (to gaseous form) in an ionic solid. Ionics that contain ions with the smallest radii and largest charge will have the largest lattice energy (the configuration of the ions is also a consideration, but we won't be asked about lattice energy at that level of detail on the AP Chemistry exam). Larger charges create a greater force of attraction. For smaller ions remember that the force of the electric field produced by a charged particle is inversely related to the square of the distance from that particle. Ions with smaller radii will be closer to the ions they are attracting and will thus exert a greater force on them, so the lattice energies of the answer choices are arranged from highest lattice energy to lowest. Since chlorine is the common anion to all of them, only the size and charge of the cation is needed to answer the question.

$FeCl_3 > MgCl_2 > CaCl_2 > NaCl > KCl$

53. (E) Only covalent compounds use prefixes such as di- and tri-. Since any binary compound containing a metal and a nonmetal is ionic, we can eliminate choices (A), (B), and (C). The key in naming this ionic is to correctly name the anion, N^{3-}. Because it is an anion we can eliminate choice (C) (again). Since there are no oxygen atoms involved, its name can't be something that ends in -ate (or -ite). The -ide suffix is a general suffix for a monatomic anion of any magnitude of charge.

54. (C) Bond order (BO) is the number of chemical bonds between a pair of atoms. It is part of molecular orbital (MO) theory, another model of bonding. (Lewis-dot structures and VSEPR are other systems used to model bonding and compound structures.)

Bond	Bond Order
Single	1
Double	2
Triple	3

The bond order does not have to be a whole number. We don't have to work out molecular orbitals to answer a question like this. If you do not remember the formula for bond order BO = ½ (number of bonding electrons – number of antibonding electrons), there's a much simpler way. First, we draw the Lewis structure. If there's resonance, we will likely have a fractional bond order. In a molecule of O_3, there are 18 valence electrons to account for. Either of the structures below could represent ozone.

But the real ozone molecule has equal bond lengths and strengths between the oxygen atoms.

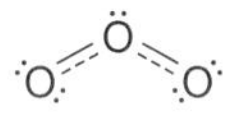

We can think of each oxygen as being bonded to another oxygen by 1½ bonds. The pi electrons (from the unhybridized p orbitals) are shared between all the oxygen atoms instead of localizing between the two atoms directly involved with the bond.

The shortcut for calculating fractional bond orders is to take the total number of bonds in one of the Lewis structures (in this case, three) and divide it by the minimum number of single bonds that would connect the atoms in question (in this case, two). For ozone, $3 \div 2 = 1.5$.

55. **(A)** N_2 has a triple bond. (See **Answer 54** for an explanation of bond order.)

56. **(C)** The ending -ide tells us that we are *not* dealing with an oxygen containing anions (which end in -ate or -ite, like phosphate, PO_4^{3-}), just a lone P with whatever negative oxidation state it typically carries (−3). Since Cl^- has a −1 charge, we know that X carries a +2 oxidation state. Then we assign the numerical value of the oxidation state of the cation as the subscript of the anion and vice versa:

$$X^{2+} + P^{3-} \rightarrow X_3P_2$$

57. **(B)** For diamond to become graphite, covalent bonds must be broken. Graphite is a network solid composed of sheets of carbon stacked on top of one another. Within each sheet (called *graphene*), the carbon atoms are covalently bonded to three other carbon atoms. The sheets are *not* covalently bonded to each other, however. London dispersion forces keep the sheets together in stacks.

Diamond is a network solid, too, but one in which *all* the carbon atoms are bonded to four other carbon atoms, so every carbon atom is covalently linked to its partners. (See **Answer 94** for a comparison of the structures of diamond and graphite.)

Choices (A) and (C) are phase changes, while (D) and (E) are ionic compounds dissolving in solution.

58. **(D)** See **Answer 57**. The sublimation of carbon dioxide differs from that of graphite because there are only van der Waals forces holding the individual carbon dioxide molecules together in the solid, and only those forces are broken when CO_2 sublimes.

59. **(E)** See **Answers 57 and 94** for descriptions of diamond structure and bonding.

Orbital Hybridization "Cheat Sheet"

Hybrid	Orbitals Involved (for n = 2 or 3)	Orbitals Left Unhybridized	Electron Geometry	Bonding
sp^3	1 s + 3 p	none	Tetrahedral	σ only
sp^2	1 s + 2 p	p	Trigonal planar	σ and π
sp	1 s + 1 p	2 p	Linear	σ and 2 π
sp^3d^2	1 s + 3 p + 2 d	3 d	Octahedral	σ only
sp^3d	1 s + 3 p + 1 d	4 d	Trigonal bipyamidal	σ only

60. **(C)** The nitrogen in ammonia has a tetrahedral electron geometry. (See the orbital hybridization cheat sheet below **Answer 59**.) Since there are only three hydrogen atoms bonded to it, a single unbonded pair of electrons remains on the nitrogen atom, making the molecular geometry of ammonia trigonal pyramidal. This asymmetry greatly increases its polarity.

61. **(A)** Compounds with trigonal bipyramidal and octahedral electron geometries have orbital hybrids that contain one or two d orbitals, respectively. Molecules with trigonal bipyramidal molecular geometry have five atoms bound to their central atom. (Though not all molecules with five atoms bound to a central atom are trigonal bipyramidal. BrF_5, for example, is square pyramidal.) Whereas compounds with octahedral molecular geometry have six atoms bound to their central atom. (See the orbital hybridization cheat sheet below **Answer 59**.)

Phosphorus and sulfur are important exceptions to the octet rule (and anything after chlorine on the periodic table may, but doesn't have to, obey the octet rule). Phosphorus can have up to five electron domains (easy to remember because the "ph" in phosphorous has the f sound, like five) and **s**ulfur can form up to **s**ix. Boron (see **Answer 62**) and hydrogen are two other exceptions.

62. **(B)** Boron is an important exception to the octet rule. We can remember the mnemonic device "B is happy with 3" because boron can form three bonds with other atoms without an unbonded electron pair on the boron, so molecules like BH_3 and BF_3 are trigonal planar in both electron and molecular geometry. (See the orbital hybridization cheat sheet below **Answer 59**.) Boron *can* form four bonds, but it requires one of the atoms to contribute both electrons to the bond without an electron contribution from boron (forming a coordinate covalent bond). (See **Answer 334** for a description of coordinate covalent bonds.)

63. **(E)** The S—O bonds in sulfur dioxide have a bond order of 1.5. The sulfur atom is sp^2 hybridized, but is bonded to only two other atoms, so the third hybrid orbital has a lone pair of unbonded electrons, repelling the oxygen atoms. Remember that electron domains for nonbonding electron pairs exert a greater force on neighboring electron domains than bonded electron domains. (See the orbital hybridization cheat sheet below **Answer 59**, and see **Answer 54** for an explanation of bond order and resonance. **Answer 84** explains the effect of unbonded electron pairs on bond angles.)

64. **(D)** The structure of carbon dioxide is linear: O=C=O. Each oxygen atom has two unbonded electron pairs. The C=O bonds are polar, but the symmetry of the molecule make it nonpolar. (See the orbital hybridization cheat sheet below **Answer 59**.)

65. **(C)** See **Answer 60**.

66. **(C)** Dipoles occur in molecules due to a nonuniform distribution of charges in the molecule. Typically, this occurs because the electron density is not equally shared between atoms. The O—H bonds in water are very polar. The electron density is greater around the oxygen atom compared with the hydrogen atom. The oxygen atom also has two unbonded pairs of electrons, which makes the water molecule bent. This makes the dipole moment of water 1.85 debye.

67. (B) Pi (π) bonds are covalent bonds involving the overlap of the two lobes of an unhybridized p orbital. The electron overlap occurs above and below the plane of the nuclei of the two atoms involved, but does not occur between the two nuclei (as in a sigma, σ, bond). Only double and triple bonds involve π bonds. All bond orders contain a σ bond. A single bond is simply a σ bond, a double bond consists of one σ bond and one π bond, and a triple bond consists of one σ and two π bonds. The molecule with the greatest number of π bonds is the one with the most double and triple bonds. C_6H_6, benzene, has three double bonds (although the bond order of the C—C bond in benzene is really 1.5 due to resonance and delocalized π electrons, see **Answer 67** for an explanation of the bond order and resonance in benzene).

68. (B) A sigma (σ) bond is one in which the region of electron overlap between the atoms in the bond is between, and in the same plane as, the two nuclei. All bonds contain *one* σ bond, but double and triple bonds also contain an addition one or two pi (π) bonds, respectively. (See **Answer 67** for a description of π bonds.)

69. (D) An atom that has sp hybridization will have two sp hybrid orbitals and two unhybridized p orbitals. A molecule with an sp hybridized central atom will most likely be linear, at least with respect to that part of the molecule (see the orbital hybridization cheat sheet below **Answer 59**).The unhybridized p orbitals don't have to contain electrons, but if they do, they are likely to be involved with pi bonds. Since we have two unhybridized p orbitals, we can form two pi bonds, either two double bonds, or one triple bond. The carbon in CO_2 is sp hybridized, which allows it to form two double bonds with each oxygen atom. (**Answers 67 and 68** describe pi and sigma bonds, respectively.)

70. (E) CH_2O is the empirical formula for a monosaccharide, but it also the molecular formula for formaldehyde (its systematic name is methanal). Because there is exactly *one* double bond (the carbonyl carbon and the oxygen), there is exactly *one* pi bond. (See **Answer 67** for a description of pi bonds.)

71. (D) Hydrogen fluoride (HF) has the largest dipole moment of the molecules listed because the electronegativity difference between two atoms (H and F) is the largest. PH_3 is a polar compound, but its dipole moment is less (0.58 μ, or debye, a coulomb meter) whereas the dipole moment of HF is a whopping 1.91 μ. The dipole moment is calculated as the product of the magnitude of the charges (or partial charges) and the distance between them. Values range from about 0–11 μ.

72. (A) Carbon monoxide, CO, has a triple bond. Double and triple bonds contain π bonds. A double bond consists of one σ (sigma) and one π bond, whereas a triple bond contains one σ and *two* π bonds. CO_2, with its two double bonds, would have also been correct had it been an answer choice. (**Answers 67 and 68** describe pi and sigma bonds, respectively.)

73. (E) A combustion reaction is a self-propagating exothermic reaction that combines oxygen with a substance and produces an oxide of the element.

74. (C) PH_3 (phosphane, also called phosphine) has a tetrahedral electron geometry, but since only three hydrogen atoms are bound, there is a single unbonded pair of electrons on the phosphorous atom, making it trigonal pyramidal in molecular geometry. (See the orbital hybridization cheat sheet below **Answer 59**.)

75. (B) All single bonds are sigma bonds, as well as one of the bonds in a double bond and one of the bonds in a triple bond. C_2H_4 has six atoms, so it's a good place to start looking for the most sigma bonds. In fact, it has five: one of each of the C—H bonds (four in all) plus one of the two bonds in the double bond between the carbons. (**Answers 67 and 68** describe pi and sigma bonds, respectively.)

76. (E) Allotropes are pure forms of the same element with different structures. Atmospheric oxygen (O_2) and ozone (O_3) are well known allotropes of oxygen, as graphite and diamond are famous carbon allotropes.

77. (B) The carbon in CO_2 is double bonded to each of the oxygen atoms.

78. (C) When looking for the compound with the greatest dipole moment, look for highly polar bonds. The H–O bond in water is the most polar bond of the compounds listed. The electronegativity difference between H–O is larger than H–N, so its dipole moment is greater. (See **Answer 71** for an explanation of how dipole moments are calculated).

79. (D) See **Answer 60**.

80. (E) The electron geometry around the nitrogen atom in NH_3 is tetrahedral (see the orbital hybridization cheat sheet below **Answer 59**), but because there are only three atoms bound to the nitrogen, there remains an unbonded pair of electrons. Therefore, the molecular geometry of NH_3 is trigonal pyramidal. (See **Answer 84** for an explanation of the effect of unbonded electron pairs on molecular geometry.)

81. (C) Ethane is an alkane (a saturated hydrocarbon) with the formula C_2H_6. Both carbons are sp^3 hybridized. (See the orbital hybridization cheat sheet below **Answer 59**.)

82. (E) Hexene is an alkene with the formula C_6H_{12}. It has one C=C double bond. The carbons *not involved* in the double bond are sp^3 hybridized. The two carbons involved in the double bond are sp^2 hybridized. (See the orbital hybridization cheat sheet below **Answer 59**.)

83. (D) Butyne is an alkyne with the formula C_4H_6. It has one C≡C triple bond. The carbons *not involved* in the triple bond are sp^3 hybridized. The two carbons involved in the triple bond are sp hybridized. (See the orbital hybridization cheat sheet below **Answer 59**.)

84. (C) Electron domains for nonbonding electron pairs exert a greater force on neighboring electron domains than bonded electron domains. CCl_4 has both tetrahedral electron geometry and molecular geometry (see the orbital hybridization cheat sheet below **Answer 59**). When the molecular and electron geometries are the same, each electron domain has an atom bound to it. The 109.5° bond angle agrees perfectly with the angles representing the four corners of a tetrahedron.

The PCl_3 molecule has a tetrahedral electron geometry but a trigonal pyramidal molecular geometry because there is one pair of unbonded electrons on the central phosphorous atom. The bond angles are less than they are for a tetrahedral—approximately 107° (although some have reported bond angles of around 100°, probably due to the further repulsion between the unbonded pair and the chlorine atoms).

Water has tetrahedral electron geometry but is a bent molecule because there are two unbonded electron pairs. The bond angles in the bent water molecule are 104.5°.

85. (C) The difference in electronegativity between carbon and oxygen is about 0.9 so the two C=O bonds in CO_2 are polar. However, CO_2 is linear and, therefore, the dipole moments of each of the bonds are 180° relative to each other, cancelling each other out and making CO_2 a nonpolar molecule even though it has polar bonds.

Choices (A) and (B) can be eliminated right away because they are molecules consisting of only one kind of atom, so the bond has to be nonpolar since there is no electronegativity difference between them. C—H bonds are not very polar. The electronegativity difference between carbon and hydrogen is about 0.35. The molecule in choice (E), difluoromethane, has two polar bonds. The electronegativity difference between carbon and fluorine is about 1.4, but the electrons are not distributed evenly around the molecule (of course they are moving but they still form dipoles due to the presence of atoms with greater electron-drawing power in the molecules). This makes CH_2F_2 a polar molecule with polar bonds.

86. (C) This question is most easily answered by knowing the formula for the alkanes and the aldehydes and carboxylic acid functional groups. The first three compounds are all hydrocarbons. The formula for the alkanes (the series of hydrocarbons with only C—C single bonds in which every carbon is completely saturated with hydrogen) is C_nH_{2n+2}. Choice (C) is the only hydrocarbon that fits into the formula. Choice (D) is acetyladehyde. An aldehyde group is a carbon double bonded to an oxygen (C=O) that occurs on the first (or last) carbon of a compound. We can recognize an aldehydes group in a chemical formula by the appearance of CHO. Choice (E) is propanic acid. The COOH in the chemical formula indicates the presence of a carboxyl group (which contains an oxygen double bonded to a carbon, and a hydroxyl group bonded to the same carbon).

87. (D) The PCl_5 molecule consists of one phosphorus atom bonded to five chlorine atoms with no unbonded electron pairs around the phosphorus (remember the mnemonic: "Phosphorus can form Five" bonds). There are five sp^3d hybrid orbitals (the "averaging" of an s, three p orbitals, and one d orbital for a total of five hybrids). (See the orbital hybridization cheat sheet below **Answer 59**.)

88. (C) The least polar bond will contain the two atoms with the most similar electronegativities, fluorine and oxygen. The smaller the electronegativity difference between the two atoms in the bond, the more uniformly distributed the electron cloud is shared between them.

89. (A) Ozone has a bond order of 1.5, which means that instead of one single and one double bond, the central oxygen is bonded to the two outer oxygen atoms by a bond of intermediate strength and length. (See **Answer 54** for an explanation of bond order and resonance in ozone.)

90. (D) Dipoles occur in molecules due to nonuniform distribution of charges in the molecule. Typically, this occurs because the electron density is not equally shared between atoms. The N—H bonds in ammonia are very polar, as the electron density is greater around the nitrogen atom compared with the hydrogen atom. The nitrogen atom also has an unbonded pair of electrons, which makes the ammonia molecule trigonal pyramidal in shape. The dipole moment of ammonia is 1.47 debye.

Chapter 3: States of Matter

91. (A) Gold (Au) is a metal, a lattice of cations bathing in a sea of electrons. (See **Answer 92** to contrast metal and ionic lattices.)

92. (B) $MgCl_2$ is an ionic lattice that contains both cations and anions. An *ionic lattice* has very different properties compared to the lattice in metals. A lattice describes a structure. The lattices in ionic compounds have roughly the same characteristics—alternating positive and negative ions—though the actual patterns of the lattices vary. Ionic lattices are very strong and rigid. They are not malleable or ductile, and they are poor conductors of heat and electricity because, unlike metals, the charges in ionic lattices are not free to roam. In other words, their charges are not mobile. The positive and negative ions are perfectly positioned to have maximum stability. Except for the ever-present vibration of atoms (at temperatures above 0 K), there's nothing moving in the lattice. The lattice structure of a *metal*, however, has only positive charges. Like an ionic lattice, the cations in a metal are positioned very regularly, but they are not as rigid. Metals are malleable and ductile partly because their electrons *are* free to roam, particularly in a wire in which a current is applied, but mostly because the metal's cations are able to take new positions relative to each other in the lattice (when a stress is applied) without breaking the metallic bonds.

93. (E) Carbon dioxide exists as individual molecules. The carbon is double bonded (a double bond consists of one σ [sigma] and one π [pi] bond; see **Answers 67 and 68** for descriptions of pi and sigma bonds, respectively) to the carbon atom (sp hybridization on the C; see the orbital hybridization cheat sheet below **Answer 59**). Carbon dioxide forms a solid only under high pressure and very low temperatures mainly due to slight dipoles between C and O atoms (but the symmetry of the molecule negates the polarity of the bonds) and London dispersion forces. These two types of intermolecular forces of attraction hold carbon dioxide molecules together in a solid. (See the phase diagram for carbon dioxide above **Question 99**.)

94. (D) Graphite is a network solid (like quartz and diamond) that is composed of sheets of carbon. Within a sheet (called graphene), the carbon atoms are covalently bonded to three other carbon atoms, so the hybridization of carbon in graphite is sp^2 (see the orbital hybridization cheat sheet below **Answer 59**). The electrons in the unhybridized p orbitals are *delocalized*; they spread out over several carbon atoms, creating a structure that can exert fairly strong London dispersion forces. This is what allows all the individual sheets to stick together. Graphite is brittle because even though the dispersion forces created by the pi electrons are strong, they are weak relative to covalent and ionic bonding, which hold most solids together under standard conditions. Graphite is remarkable in that it is a nonmetal solid that conducts electricity (due to the delocalized electrons, which are mobile). Like a metal, graphite has luster, but like a nonmetal, it's not malleable. It is soft and flexible, but inelastic (it doesn't reform after being deformed). Graphite and diamond are *allotropes*, pure forms of the same element with different structures. (See **Answer 95** to compare with diamond.)

95. (C) Diamond is a network solid (like quartz and graphite). In diamond, *all* the carbon atoms are bonded to four other carbon atoms, so every carbon atom is covalently linked to its partner. The carbon atoms are sp^3 hybridized (see the orbital hybridization cheat sheet below **Answer 59**), so there are no delocalized pi electrons. Diamond and graphite are *allotropes*, pure forms of the same element with different structures. (See **Answer 94** to compare with graphite.)

96. The normal boiling point is the point at which a liquid turns to a gas at an atmospheric pressure of 1 atm. On this graph, it is the point on the line between gas and liquid phases that would meet with a line drawn perpendicular to the y-axis at 1 atm. A liquid (in an open container) boils when the vapor pressure above the liquid reaches the pressure atmosphere. Since the boiling point is determined in part by the atmospheric pressure, the *normal boiling point* (at 1 atm), is the boiling point *at 1 atm.*

97. The solid area of the graph meets directly with the gas area of the graph at low temperatures and pressures, to the left of the triple point.

98. (E) Typically, high pressure favors the formation of a solid. The negative slope of the line in the phase diagram indicates that for this compound, a decreased temperature is needed for the solid to form at higher pressures. This suggests that the solid form of this compound is *less dense* than the liquid form. Water is a compound whose solid is almost always less dense than the liquid, though it actually depends on the way the crystals form (there are high and very high density forms of amorphous ice, but it's highly unlikely we'll be asked about them for the AP Chemistry exam).

The graph does not support choice (E) because the area for solids extends well below 1 atm. The line indicates equilibrium between the two phases, but the area for solids shows that the lower the temperature, the less pressure needed. A lower temperature is needed for a high pressure solidification, but a low temperature solidification (say, −100°C) requires very little pressure. (See **Answer 11** for an *except* question strategy.)

99. (E) At a constant pressure of 1 atm, the solid CO_2 sublimes directly into a gas without going through the liquid phase. To liquefy, CO_2 requires a pressure greater than 5 atm. At temperatures below −56°C approximately, CO_2 does not liquefy at all.

100. (B) See **Answer 99**.

101. (D) We should immediately recognize *SiO_2 as a network solid* (silicon dioxide, also known as quartz). *Network solids, like diamond and graphite, have high melting points.* SiO_2 is not a molecular formula, it is an empirical formula that represents the ratio of Si to O atoms in the compound. The melting point of SiO_2 is ~1,600–1,725°C. H_2S and C_5H_{12} are both gases under standard conditions. I_2 and S_8 are solids but their melting points are low, relative to SiO_2, at ~114°C and 115°C, respectively.

102. (D) This information is obtained directly from the graph. One kilogram of water absorbed approximately 2,300 kJ of energy to vaporize.

103. (B) Water is different in that the density of the liquid is *greater* than that of the solid (why ice floats). Each water molecule in solid water (ice) is connected to four other water molecules by hydrogen bonds. This spreads the molecules further apart than when they are in a liquid, where each water molecule is hydrogen bonded to two or three other water molecules at a time.

104. (C) During fusion (melting), the temperature of water doesn't change. Use the formula $q = mH_{fus}$, solving for H_{fus}. There are approximately 350 kJ of heat absorbed by 1 kg of water during the interval where the water is at 0°C (the temperature, or average kinetic energy, of a substance does not change while it is undergoing a phase change, see **Answer 109** for an explanation).

105. (D) The 1.0 kg water absorbed approximately 2,300 kJ to vaporize.

106. (A) There is a temperature change, so we use the formula $q = mc\Delta T$ and solve for c, the specific heat $\therefore\ c = {}^{q}/_{m\Delta T} = {}^{100\text{ kJ}}/_{(1.0\text{ kg})(50°\text{C})} = \sim 2.0\text{ kJ kg}^{-1}°\text{C}^{-1}$.

107. (E) The heat of vaporization is much higher than the heat of fusion. The table below shows that an average of 1.5 mol of hydrogen bonds per mol of water are broken during fusion (4 mol H bonds per mol ice – average of 2.5 mol H bonds per mol water = 1.5 mol H bonds broken during fusion), whereas an average of 2.5 mol of hydrogen bonds are broken (per mol of water) during vaporization. Water molecules *do* move closer together during fusion and do move farther apart during vaporization, but this fact alone does not explain the difference in energy requirements between fusion and vaporization.

Solid (ice)	→	Liquid	→	Vapor
4 hydrogen bonds per water molecule		2–3 hydrogen bonds per water molecule		No hydrogen bonds

108. (B) With the information in the table above, the enthalpy of hydrogen bond formation can be calculated (it will have the same magnitude but opposite sign of the enthalpy of *breaking* hydrogen bonds).

Water is not forming from H_2 and O_2 gas, it is simply changing state, so we cannot calculate the enthalpy of formation from the data. Superheated steam is not represented in the graph so we have no data to use in a calculation. There is no time component, so we do not know the rate at which heat is being added, and there is no volume data with which to calculate density (density = ${}^{\text{mass}}/_{\text{volume}}$).

109. (D) The absolute temperature of a substance is proportional to the average kinetic energy (KE) of the particles in the substance. It is helpful to replace the word temperature with "average kinetic energy" when solving chemistry problems. If the temperature is increasing, the average KE of the particles is increasing. During phase changes, the temperature remains constant. Therefore, there is no change in KE. The heat added is used to change (increase) the *potential energy (PE)* of the particles, which are changing positions relative to each other during phase changes. The definition of zero entropy is a perfect, pure crystalline solid at 0 K (it's also the third law of thermodynamics). Any deviation from 0 K and/or a pure, perfect crystalline solid indicates that entropy is increasing. An increase in temperature typically increases the entropy (the units of which are J mol^{-1} K^{-1}). (See **Answer 11** for an *except* question strategy.)

110. (A) Diethyl ether. A high vapor pressure indicates the particles of the liquid are not strongly attracted to each other since a low temperature (average kinetic energy) allows them to escape from solution.

111. (D) Only substance D and E can form hydrogen bonds (because of their O—H groups) but E (methanol) has a higher vapor pressure so its intermolecular forces must be weaker. All molecules exhibit London dispersion forces, but these are very weak relative to hydrogen bonds and dipole–dipole attractions, so they are not considered significant in polar molecules. The strength of London dispersion forces increases with the number of electrons, which is typically proportional to molar mass. Particles with a high molar mass (and therefore lots of electrons) will exhibit stronger London dispersion forces. Because substance D, ethanol, is larger than methanol, it exhibits greater dispersion forces and therefore has a lower vapor pressure.

112. (C) The nonpolar compounds in the table are A (diethyl ether), B (carbon disulfide) and C (carbon tetrachloride). A volatile compound readily evaporates, will have a high vapor pressure, and have weak intermolecular forces of attraction.

113. (D) Ionic compounds that contain ions with the smallest radii and largest charge will have the highest melting points (and largest lattice energies, see **Answer 52** for an explanation of lattice energy). The configuration of the ions within the lattice is also a consideration, but we don't need to be concerned with that level of detail for the AP Chemistry exam. Larger charges on the ion create a greater force of attraction, increasing the melting point. The force of the electric field produced by a charged particle is inversely related to the square of the distance from that particle, so ions with smaller radii will be closer to the ions they are attracting and will thus exert a greater force on them, thereby increasing the melting point.

114. (E) Atmospheric pressure is the column of air above a particular area. The higher the altitude, the shorter the column of air, so the lower the pressure. The pressure drop with increasing altitude is fairly linear until about 10 km above sea level, after which it drops precipitously. Therefore, the column of air is densest closest to sea level (since the weight of the column of air above the air closest to the ground is greatest, pressing all the particles closer together). The column of air above an open container pushes down on the particles in the container that are trying to escape. The greater the pressure, the greater the escape velocity required by the particles, therefore, the higher the temperature (average kinetic energy, KE) required for escape ($KE = \frac{1}{2} mv^2$, where m = mass and v = velocity). Choice (A) is not correct because at the same temperature, the particles of any two substances have the same average kinetic energy.

115. (C) See **Answer 114**.

116. (D) Carbon dioxide exists as individual molecules. The carbon is double bonded to each of the carbon atoms in a linear molecule. Carbon dioxide forms a solid only under high pressure and very low temperatures mainly due to London dispersion forces and the slight polarity of the C and O bond (but the symmetry of the molecule mostly negates the polarity of the bonds). These two types of intermolecular forces of attraction hold carbon dioxide molecules together in a solid. Remember that *phase changes are physical*, not chemical, meaning that only intermolecular forces of attraction (IMFs, also called van der Waals forces) are being formed or broken.

Although deposition (and solidification) typically requires nucleation sites, that choice is not the best answer because it doesn't address the changes in the attractive forces. (See the phase diagram for carbon dioxide above **Question 99**.)

117. (A) Phase changes are physical changes, not chemical changes, so covalent and ionic bonds are not broken or formed. Only van der Waals forces (intermolecular forces of attraction, or IMFs) are being formed or broken. The density of liquid water is greater than that of solid water. Water, unlike most substances, is *less dense* as a solid because there are four hydrogen bonds per water molecule in ice (compared to the two to three hydrogen bonds per water molecule in liquid water) that cause the water molecules to spread farther apart from each other, forming a well-organized crystal. Because fewer molecules of water are present per volume of water in ice, the density is lower (so the solid form floats in its liquid form). (See the phase diagram for water above **Question 96**, and see **Answer 11** for an *except* question strategy.)

118. (A) The strength of London dispersion forces correlates with molar mass, but only because the number of electrons is correlated with molar mass. Helium atoms have only two electrons, so they cannot form a strong temporary dipole. Xenon atoms have 54 electrons, so they can form significant dipoles (at low temperatures). The importance of the trend in electron number and London dispersion strength is reflected in the different boiling points of He and Xe. Helium has the lowest boiling point of the elements, −269°C (a mere 4 K), while the boiling point of Xe is −108°C (165 K).

119. (B) All of the elements are nonpolar, so the only IMF to consider is London dispersion. Br_2 has the highest molar mass (and the greatest number of electrons) of the choices listed, therefore it has the strongest dispersion forces. Br_2 is also the only liquid among gases (under standard conditions), so that fact alone would indicate the highest boiling point (since the rest boiled at temperatures below 25°C if they are gases at room temperature). (See **Answers 111 and 118** for more on the relationship between London dispersion forces, molar mass, and number of electrons.)

120. (E) The question basically is asking us what happens during melting, when a substance is at its melting point and has already started but has not yet melted completely. The heating curve of water is shown above **Question 102**. Melting occurs at 0°C and evaporation occurs at 100°C. The temperature does not change during the phase changes even though the water is still absorbing heat. These temperature plateaus are due to an increased potential energy of the substance changing phases. Temperature is the average kinetic energy of the particles in a substance, so if the temperature is not changing, neither is the average kinetic energy. Covalent bonds are typically *not* broken during melting, which is a physical, not a chemical, change. The volume of a substance *often* increases with melting, since the solid form of most substances is denser than the liquid form. Water, however, is an important exception. (See **Answer 109** for a comparison of kinetic and potential energy changes that occur during heating and phase changes and **Answer 117** for a comparison of the densities of liquid and solid water.)

121. (B) The gas with the greatest mass will have the greatest density since at the same temperature and pressure, equal volumes of gas contain the same number of gas particles. One mol Xe at STP occupies 22.4 L and has a mass of 131 g ($^{131\text{ g}}/_{22.4\text{ L}} = {}^{5.85\text{ g}}/_{\text{L}}$). Helium, on the other hand, has a density of $^{0.18\text{ g}}/_{\text{L}}$ ($^{4\text{ g per mole}}/_{22.4\text{ L}}$) at STP. As long as we compare the same volumes at the same temperature and pressure, we only have to compare molar masses to compare densities.

122. (E) The absolute (Kelvin) temperature of a substance is directly proportional to the average kinetic energy (KE) of its particles. The equation for KE is ½ mv^2. Since the molar mass of compound is an intrinsic property, it remains constant. Only the velocity of the particles changes when the kinetic energy changes. Mass and velocity are both proportional to KE, but inversely related to each other. A more massive gas will move slower than a lighter one at the same temperature. The molar mass of N_2 is 28 g mol^{-1}, so any gas of similar molar mass (CO) will have a similar velocity under the same conditions. (See the **Answer 123** for the formula to calculate a ratio of gas speeds at the same temperature.)

123. (A) Effusion is the diffusion of a gas through a tiny hole or opening. The faster a gas particle can move, the more quickly it effuses (and diffuses). The speed of a particle of gas is related to its kinetic energy and molar mass.

Under the same conditions, the least massive gas (He, in this case) will effuse the fastest and the most massive gas (Xe, in this case) will effuse the slowest.

Absolute temperature is directly proportional to the average kinetic energy (KE) of the particles in a substance. KE = ½ mv^2. For the same KE, $v = \sqrt{^2/_m}$, or, we can say that the velocity is inversely proportional to the square root of the molar mass. Graham's law of effusion allows us to calculate the relative speeds of two different gases at the same temperature. To compare Xe and He:

$$\text{Rate of } {}^{\text{He}}/_{\text{Xe}} = \sqrt{(^{\text{molar mass Xe}}/_{\text{molar mass He}})} = \sqrt{(^{131}/_4)} = 5.7$$

That means that the rate of effusion of helium is 5.7 times greater than that of xenon. (See **Answer 237** for another derivation of Graham's law of effusion.)

124. (B) See **Answer 123**.

125. (A) Helium, which has the *fewest* IMFs, will require high pressures and low temperatures to liquefy. It has the lowest boiling point of all the elements, 4 K. (See **Answer 118** for a comparison of the boiling points of the noble gases.)

126. (B) Since all the gases listed are nonpolar, the only significant IMF they can form is London dispersion. The gas with the greatest number of electrons (the most massive gas) will generate the greatest London dispersion forces under the same conditions. Helium, then, will have the lowest boiling point and require the most pressure to condense into a liquid. (See **Answer 118** for more on the relationship between London dispersion forces, molar mass, and number of electrons.) Helium has the lowest boiling point of the elements, −269°C at 1 atm (a mere 4 K).

127. (B) The highest condensation temperature is also the highest boiling point, so we are looking for the gas capable of forming the strongest IMFs. A pressure of 10 atm is very high (for comparison, the air pressure in car tires is about 2 atm), but high pressure makes IMFs more likely to form, and therefore helps the gas to condense. The boiling point of Xe is −108°C at 1 atm (165 K). (See **Answer 118** for more on the relationship between London dispersion forces, molar mass and number of electrons.)

128. (D) An ideal gas is an imaginary gas whose behavior with regards to temperature, pressure, and volume is completely described by the ideal gas equation, PV = nRT (it may be helpful to remember the name "**pivnert**"). Actual measurements of T, P, and V of real gases vary (very) slightly from the predictions made by the ideal gas law. These real gases deviate mainly because of the forces of attraction and repulsion between the particles. Low pressures maintain a low density of the gas, so particles are far enough apart that attractive and repulsive forces aren't felt, and high temperatures overcome forces of attraction and repulsion. With these forces of attraction and repulsion minimized, real gases behave quite ideally.

129. (D) The two gases are at the same temperature and pressure, so equal volumes of the two gases will contain the same number of particles, but there are more CO_2 molecules (and therefore a larger volume) than O_2 molecules. CO_2 is more massive than O_2, and there are a greater number of particles, so the masses of the two gases are certainly not the same. The density of two gases at the same temperature and pressure can be compared simply by comparing their molar masses. CO_2 is more massive than O_2, at 298 K and 1 atm, the CO_2 gas is denser. If they are at the same temperature, their average KE is the same, but CO_2 is more massive and so the average velocity of the particles is less than that of the O_2 particles. Two gases at the same temperature will have the same average kinetic energy, but the more massive gas will have, on average, slower moving particles. (See **Answer 123** for an explanation of KE and average molecular, as well as Graham's law, a simple formula to calculate the relative speeds of two gases.)

130. (C) Two gases at the same temperature will have the same average kinetic energy, but the larger gas will have, on average, slower moving particles. Since these two gases are equally massive (44 g mol^{-1}), their particles will have the same average speed. The $N_2O_{(g)}$ sample has more moles so it has more particles, and it will occupy a greater volume than CO_2 at the same temperature and pressure. Notice density was *not* given as a choice. Gases of the same molar mass at the same temperature and pressure have the same densities. (See related **Question and Answer 129**.)

131. (D) This is a simple conversion. Thirty-two grams of O_2 is one mole of O_2. At 1 atm and 298 K, this number of oxygen molecules would occupy 22.4 L, but the pressure is four times. However, temperature and pressure are *directly proportional*, so a fourfold increase in pressure would be accompanied by a fourfold increase in temperature ($298 \times 4 = 950$).

We will arrive at the same answer using the ideal gas law (PV = nRT, known as **pivnert**). Or, since we know there is 1.0 mole of gas (and at 298 K and 1 atm, it will occupy 22.4 L), we can use Gay–Lussac's law: ${}^{P1}/_{T1} = {}^{P2}/_{T2}$, ${}^{1\ atm}/_{298\ K} = {}^{4\ atm}/_{T2}$ ∴ $T_2 = 950$ K. *Remember to only use the Kelvin (absolute) temperature scale when dealing with gases because there are no negative numbers.*

132. (B) Rearranging PV = nRT for P gives $P = {}^{nRT}/_{V}$. Choice (E) is incorrect because the numerical value of the gas constant depends on the units. The number 0.0821 is used when P is measured in atm. The value 8.314 is used when P is given in kPa (kilopascals). *Remember to only use the Kelvin (absolute) temperature scale when dealing with gases (because there are no negative numbers).*

133. (B) The total initial pressure in the container is the sum of the partial pressures of all the gases present (Dalton's law of partial pressures), so the total $P_{initial} = 1.2 + 3.8 = 5$ atm. Since the pressure is proportional only to number of particles at a given volume and temperature, we know that there are three times more H_2 molecules than N_2 molecules ($1.2 \times 3 = 3.8$). This ratio of H_2 to N_2 is exactly the same as the ratio of coefficients in the balanced equation, so this is not complicated. When the partial pressure of N_2 falls to 0.9 atm, that means one-fourth of the N_2 has been consumed by the reaction (0.3 is 25 percent of 1.2). If two NH_3 are formed for every N_2 molecule that is consumed, then 0.6 atm (0.3×2) of NH_3 will be formed. The important thing to remember is that the partial pressures of the gases in a mixture tells us the relative number of particles of each gas in the mixture.

134. (E) Sulfur dioxide, SO_2, has bonds of high polarity and has a bent molecular geometry, making it very polar and therefore subject to dipole–dipole IMFs and of course, London dispersion forces. Ideal gases are the imaginary gases in which the particles experience no forces of attraction or repulsion. (See **Answer 128** for a definition of ideal gases and the conditions under which real gases behave most like ideal gases.)

135. (D) See **Answer 128**.

136. (B) Dalton's law of partial pressure states that the pressure exerted by a specific gas within a mixture is proportional to the mole fraction of that gas. The trick to answering this question is realizing that if the equal masses of neon and argon are in the container, then twice the number of neon atoms are present (since the molar mass of argon is twice that of neon). Since we only need to be concerned about mol fraction and no other information is given, let's simply assume we have 1 mole Ar (40 g) and 2 moles Ne (40 g). That means there's a total of 3 mol of gas, of which one-third are Ar atoms and two-thirds are Ne atoms. If the total pressure is 1.2 atm, one-third of that pressure, or 0.4 atm, is due to Ar and 0.8 atm is due to Ne. It doesn't matter what *actual* number of mole we use, only the ratio between Ar and Ne. As long as there are twice as many Ne atoms as Ar atoms in our calculations, our answer will be the same (and correct).

137. (E) Dalton's law of partial pressure states that the pressure exerted by a specific gas within a mixture is proportional to the mole fraction of that gas. The total number of moles of gas is $0.5 + 1 + 1 = 2.5$ mol. The pressure exerted by each mole of gas $= {}^{750\ mmHg}/_{2.5\ mol} =$ 300 mmHg per mole gas. If a half mole of $SO_{2(g)}$ is present, then half of 300 mmHg, or 150 mmHg, of pressure is exerted. We will arrive at the same answer by calculating the mole fraction of $SO_{2(g)}$ (${}^{0.5\ mol}/_{2.5\ mol} = 0.2$) and multiplying it by the total pressure (0.2×750 mmHg $= 150$ mmHg).

138. (D) A 2-L container will hold approximately 0.1 mole of gas at STP. The molar mass of O_2 is 32 g mol^{-1} $\therefore$ 0.1 mol = ~3 g.

139. (C) The 2-L flask holds a total of three mole gases at a pressure of 800 mmHg. Dalton's law of partial pressure states that the pressure exerted by a specific gas within a mixture is proportional to the mole fraction of that gas. The pressure of each gas can be calculated as follows. We can easily check our work because the sum of the partial pressures should equal the total pressure.

Moles Specific Gas	÷ Total Mole Gas = Mole Fraction	× Total Pressure = Partial Pressure
0.50 mole SO_2	0.17	136
0.75 mole of O_2	0.25	200
0.75 mole of CH_4	0.25	200
1.00 mole of CO_2	0.33	264

140. (D) This is a Graham's law of effusion problem. The two gases are released at opposite sides of a tube and if they diffused at the same rate, they would meet in the middle. The speed at which a gas diffuses at a given temperature is inversely proportional to the square root of its molar mass (see **Answer 123** for a derivation of Graham's law). The larger mass of HCl makes it diffuse more slowly than NH_3 by the following equation $\text{rate } NH_3/\text{rate HCl} = \sqrt{36.5/17}$ = NH_3 diffuses 1.5 times as quickly, therefore in the same period of time, NH_3 will travel 1.5 times as much distance.

If distance HCl travels = x, the distance NH_3 travels = 1.5x.

Total distance = $1x + 1.5x = 2.5x = 100$ cm $\therefore$ $x = 40$ cm and $1.5x = 60$ cm.

141. (D) A 2-L container will hold approximately 0.1 mole gas at STP. The molar mass of Cl_2 is 71 g mol^{-1} $\therefore$ 0.1 mole = ~7 g (**Question 138** is similar).

142. (A) This question is asking us to identify the gas that is *least soluble in water*, and therefore if collected above water, will produce a high yield because the least amount will be dissolved into the water (and therefore not collected).

143. (D) It is sometimes useful to imagine the atmosphere as a very thin, light fluid. The rubber duck in a bathtub floats because the average density of the duck is less than the density of the water as is the average density of a luxury ocean liner. A submarine, on the other hand, can change its average density to sink or float.

144. (D) The term *constant temperature* is our clue that the average kinetic energy of the particles will be the same. Although the speed of the particles is related to their kinetic energy, so is their mass. At the same temperature, more massive gases will move with slower speed. Since all these particles are the same, their speed remains the same at the same temperature.

145. (A) The speed at which a gas diffuses (or effuses through a tiny hole) is inversely proportional to the square root of its molar mass (see **Answer 123** for a derivation of Graham's law). Therefore, Ar will effuse out of the container the fastest, leaving the least number of Ar particles behind, and therefore exerting the least partial pressure. Kr is of intermediate mass between Ar and Xe and so will effuse at an intermediate speed. Kr will effuse the slowest leaving the greatest number of particles in the container and therefore having the highest partial pressure.

146. (B) The speed at which a gas diffuses (or effuses through a tiny hole) is inversely proportional to the square root of its molar mass (see **Answer 123** for a derivation of Graham's law). The lightest gas in the list is H_2, so at the same temperature (average kinetic energy), H_2 will move the fastest.

147. (A) The sodium carbonate reacts with hydrochloric acid according to the equation:

$$Na_2CO_{3(aq)} + 2\ HCl_{(aq)} \rightarrow 2\ NaCl_{(aq)} + CO_{2(g)} + H_2O_{(l)}$$

$$0.250\ L\ HCl \times {}^{2.5\ mol}/_{L} = 0.63\ mole\ HCl$$

$$10.6\ g\ Na_2CO_3 \times {}^{1\ mol}/_{106\ g} = 0.10\ mole\ Na_2CO_3$$

Remember to check for limiting reagents if the amounts of two reactants are given. A simple check is to take the number of moles of each reactant and divide it by its stoichiometric coefficient in the balanced equation.

$$0.63\ mole\ HCl \div 2 = 0.325$$

$$0.10\ mole\ Na_2CO_3 \div 1 = 0.10\ (limiting\ reactant)$$

Since there is a 1:1 ratio between Na_2CO_3 consumption and CO_2 formation, 0.11 mole CO_2 is the theoretical yield.

148. (A) Carbon dioxide, $CO_{2(g)}$, is relatively nonpolar, so not a great deal of it dissolves, but a small fraction does (from about 3.5 g CO_2 per kg water at 0°C to 0.5 g CO_2 per kg water at 60°C at sea level). Importantly, CO_2 reacts with water to form carbonic acid, a weak acid, according to the equation.

$$CO_{2(g)} + H_2O_{(l)} \Leftrightarrow H_2CO_3 \Leftrightarrow H^+_{(aq)} + HCO_3^-{}_{(aq)} \Leftrightarrow H^+_{(aq)} + CO_3^{2-}{}_{(aq)}$$

(Our lives depend on this reaction. CO_2 is transported in the blood mainly as HCO_3^-, and the regulation of our breathing relies on it. The reaction occurs in red blood cells with the help of the enzyme *carbonic anhydrase.*)

149. (D) Any gas collected over water will be a mixture of at least two gases, the gas (or gases) produced in the reaction *and* water vapor from the eudiometer. Just remember Dalton's law of partial pressures, *the total pressure of a mixture of gases is the sum of the partial pressures of its constituent gases.* The partial pressure of each gas is determined solely by its mole fraction in the mixture.

Because we don't know the mole fraction of either gas, we determine it by the pressure. We start with the partial pressure of the water and we use a handy fact: *The vapor pressure of water is determined solely by the temperature.* If the vapor pressure of water at 22°C is about 20 mmHg (this information would be provided), we can then deduce the pressure of the gas in our mixture.

The total pressure of the gas in the eudiometer is the atmospheric pressure in the lab, which was given as:

760 – 20 = 740 mmHg of $CO_{2(g)}$. Then we use the ideal gas law, PV = nRT, to calculate the number of moles of CO_2.

150. (D) First start by calculating the number of moles of each gas:

1.6 g He = 0.4 mole, 4 g Ar = 0.1 mole, 26 g Xe = 0.2 mole

Total number of moles = 0.7

P = 2.1 atm ∴ 0.3 atm per 0.1 mole of gas

0.2 mole Xe ∴ 0.6 atm of pressure

We can also find the mole fraction of Xe and multiplying it by the total pressure:

$^{0.2\text{ mol Xe}}/_{0.7\text{ total mol}} = 0.29 = \sim 0.3$

$0.3 \times 2.1\text{ atm} = 0.63 = \sim 0.6\text{ atm}$

Chapter 4: Solutions

151. (A) Raoult's law states that the vapor pressure of a solution of a nonvolatile solute is equal to product of the vapor pressure of the pure solvent and its mole fraction. A solution will have a *lower* vapor pressure than the pure solvent and therefore a higher boiling point because a liquid or solution boils when the vapor pressure above the liquid reaches the pressure of the atmosphere. Solutes will also depress the freezing point and increase the osmotic pressure in direct proportion to their concentration. (See **Answer 459** for the formula to calculate the DT of boiling points or freezing points of solutions.)

152. (A) Most substances undergo a change in density (and volume) with a change in temperature. Mass, however, doesn't change with temperature. Molality (*m*) is the concentration expressed in mole per *kilogram* solvent, a unit of mass. Molarity (M) is a unit of concentration expressed in mole per *liter*, a volume.

153. (C) We need to convert mL to L to solve for number of mole from molarity, $^{mol}/_{L}$.

$0.125\text{ L} \times (^{0.2\text{ mol}}/_{L}) = 0.025\text{ mol } CuSO_4 \cdot 5\ H_2O$.

$0.025\text{ mol} \times (^{250\text{ g}}/_{mol}) = 6.25\text{ g}$

154. (B) This is a dilution problem. Use $M_1V_1 = M_2V_2$

$$M_{HCl}V_{HCl} = M_{0.8Msol}V_{0.8Msol}$$

$$(20)\ (5) = (0.8)\ (V_{0.8Msol})$$

$$V_{0.8Msol} = 125\ mL$$

But this is the *final volume* of the solution. The question asked *how much distilled water must be added.* We must subtract the 20 mL of 5 M HCl from the final volume to calculate the volume of water added:

$$125\ ml_{FinalSol} - 20\ mL_{HCl} = 105\ mL_{water}$$

155. (E) The lead nitrate and sodium chloride react according to the equation:

$$Pb(NO_3)_2 + 2\ NaCl \rightarrow PbCl_{2(s)} + 2\ NaNO_{3(aq)}$$

$$0.100\ L\ Pb(NO_3)_2 \times (^{0.2\ mol}/_{L}) = 0.02\ mol\ Pb^{2+}$$

$$0.100\ L\ NaCl \times (^{0.3\ mol}/_{L}) = 0.03\ mol\ mol\ Cl^-$$

Since two Cl^- are needed for each Pb^{2+}, **Cl^- is our limiting reactant** (remember the simple check: divide the number of mole of each reactant by its stoichiometric coefficient in the balanced equation, the smallest quotient is the limiting reactant) and we'll have excess Pb^{2+} ions in solution.

$$0.03\ mole\ Cl^- \times (^{1\ mol\ Pb^{2+}}/_{2\ mol\ Cl}) = 0.015\ mole\ of\ Pb^{2+}\ ions\ precipitated\ out\ with\ the\ chloride\ ions$$

$$0.02\ mole\ Pb^{2+} - 0.015\ mole\ Pb^{2+}\ precipitated = 0.005\ mole\ Pb^{2+}\ excess$$

0.005 mole Pb^{2+} in a *final volume* (don't forget to add the volumes of both solutions) of 200 mL (or 0.2 L) = 0.025 M.

156. (E) The two solutions were prepared with the same number of moles of their respective compounds, so the determinant of electrical conductivity is the *concentration of ions* of their solutions. A particle that does not dissociate into ions when dissolved does not conduct electricity. The higher the ion concentration of a solution, the greater its electrical conductivity. Ions (or charges) *must be mobile* in order to conduct electricity, so *solid salts do not conduct electricity* (but molten salts do). Soluble salts and strong acids and bases are the best electrolytes because they completely dissociate in water, producing at least 2 mole ions per mole compound.

157. (C) Since we're given a percent and not an absolute mass, assume a 100 g sample of solution ∴ 66% C_2H_4O = 66 g C_2H_4O = 1.5 moles.

100 g of sample – 66 g C_2H_4O = 34 g of water, or about 1.9 (round to 2) mole water. Total moles = 3 ∴ $^{1.5\ mol\ ethanol}/_{3\ mol\ total} = 50\%$.

158. (A) There are 2 moles of ethanol in a total of 10 mol solution (20 percent).

144 g H_2O @ (18 g mol^{-1}) = 8 moles

92 g ethanol @ (46 g mol^{-1}) = 2 moles

159. (D) 29 g NaCl @ (58 g mol^{-1}) = 0.5 mole NaCl

$^{0.5 \text{ mol NaCl}}/_{0.2 \text{ kg solvent}} = 2.5$ *m* solution

160. (D) Phosphates are not particularly soluble. All nitrate and ammonium salts are soluble, as well as all group 1 salts. Acetate is also very soluble.

161. (C) The addition of HF, a weak acid, will lower the pH by increasing the H^+ concentration in the solution. The increased H^+ concentration will shift the equilibrium of the BaF_2 to favor its dissociation (and therefore its solubility) by providing the F^- already in the solution with more H^+ ions to bind to form HF. F^- is a good conjugate base, so it will take up many of the H^+ ions. The H^+ and F^- ions in the acid are already at equilibrium, so adding them to the BaF_2 solution will not increase the F^- concentration when it is added, as much as it will increase the H^+ concentration. This is because the F^- that is being added will bond to the H^+ ions to form HF, further lowering the F^- concentration, and further pulling the solubility equilibrium to the right, favoring dissociation and increased solubility. By the time the two solutions have completely mixed and a new equilibrium is achieved, the final F^- concentration is lower, allowing more Ba^{2+} ions to be dissolved.

162. (D) The formation of hydrogen bonds is *exothermic.*

163. (A) Assume 100 mL of each liquid.

100 mL ethanol × $^{0.79 \text{ g}}/_{\text{mL}}$ = 79 g ethanol @ 46 g mol^{-1} = 1.7 moles

100 mL water × $^{1.0 \text{ g}}/_{\text{mL}}$ = 100 g H_2O @ 18 g mol^{-1} = 5.5 moles

Total mol solution = 1.7 + 5.5 = 7.2 moles

$^{1.7 \text{ mol ethanol}}/_{7.2 \text{ mol total}} = 0.24$

164. (A) Different proportions of solute and solvent can produce different enthalpy changes, but the solvation of ethanol and water is unusual in that it starts out exothermic at low concentrations of water, changes to endothermic in the mid-range, and then reverts back to exothermic at high concentrations of water. Solutions are complex, but we can still arrive at a fairly simple but logical interpretation of this situation. If a particular solvation process is exothermic, then more (or "stronger") intermolecular forces of attraction (IMFs) are formed than broken (IMF formation is exothermic). If it is endothermic, then more (or stronger) IMFs are broken than formed (IMF "breakage" is endothermic).

165. (D) Any particle with electrons will exhibit some amount of London dispersion forces, so our answer must contain choice III. The name *ethanol* lets us know the compound is an alcohol, but the —OH group is obvious from the chemical formula given at the beginning of the question set. Water, of course, has two O—H bonds. Molecules with

O—H, N—H, and F—H bonds can form hydrogen bonds. *Hydrogen bonding is the strongest of the intermolecular forces of attraction* (technically, hydrogen bonding is a special case of dipole–dipole attraction). The C—H bonds in ethanol are fairly nonpolar, so there are no dipole–dipole forces holding these two molecules together.

166. (C) First, we need to determine the number of moles of Cl^- ions in the solution. There are 0.1 mole from KCl, 0.2 mole from $CaCl_2$, and 0.3 mole from $AlCl_3$ ∴ 0.6 mole Cl^- in 1-L solution. Because each Pb^{2+} ion can precipitate 2 Cl^- ions, 0.6 ÷ 2 = 0.3, the minimum number of moles of Pb^{2+} ions needed.

167. (B) The solubility of gases in water is greatest at *high pressures* and *low temperatures.*

168. (C) Water soluble salts are *not* soluble in nonpolar solvents, like CCl_4. Salts dissolve by dissociating. Their ions form interactions with solvent molecules that have polar groups on them (ion-molecule attractions). A Na^+, for example, will be attracted to the partially negatively charged oxygen atom in an O—H bond (of water, for example), whereas the Cl^- ion will be attracted to the partially positively charged hydrogen atom of the O—H bond.

169. (C) We are looking for the compound whose solubility is drastically reduced by a reduction in temperature. If this data were presented in a graph, we'd be looking for the solubility curve with the steepest slope within the temperatures given in the question.

170. (B) The Ag^+ is not soluble as a chloride or a sulfate, but it is soluble as a nitrate.

171. (A) The chloride from NaCl forms an insoluble precipitate with Ag^+ (AgCl).

172. (E) First, we calculate the number of moles of NaOH (and OH^-):

60 mL × 0.4 M = 24, *but* the unit M is mol L^{-1}, so we need to move the decimal over three places to account for 1,000 mL per liter ∴ 0.024 mol NaOH (or, convert to L before doing the calculation) ∴ 0.024 mole OH^- (only one OH^- per NaOH).

Next we calculate the Ba $(OH)_2$ (and OH^-): 40 × 0.6 M = 0.024 mol $BaOH_2$ ∴ 0.048 mole OH^- (2 moles OH^- per mole $BaOH_2$).

There are a total of 0.072 moles OH^- (0.024 + 0.048) in 100 mL (0.1 L, add the volumes) so the molarity is 0.72 M.

We can use a variation on $M_1V_1 = M_2V_2$ to solve this problem, but we need to remember that *when mixing solutions, we must add their volumes.*

$$M_1V_1 + M_2V_2 = M_3V_3$$

$$(0.06)\ (0.4) + 2\ (0.04)\ (0.6) = M_3\ (0.1)$$

173. (C) The first solution, a mixture of $CuCl_2$ and $MgSO_4$, is totally soluble. That not only tells us that $CuCl_2$ and $MgSO_4$ are soluble, but the products of the double replacement reaction between them are also soluble. That would be $CuSO_4$ and $MgCl_2$.

When the student then mixes $Al_2(SO_4)_3$ and CuF_2, however, a precipitate forms. We assume $Al_2(SO_4)_3$ and CuF_2 are soluble because they were combined as solutions, but one of the products of the double replacement is obviously not soluble. AlF_3 and $CuSO_4$ are the products. We may have our solubility rules memorized, but for this particular question, we don't need them. The first solution made by the student already tells us that $CuSO_4$ is soluble, so we must conclude it's the AlF_3.

The important thing to remember for this question is about making sure the charges on our ions are correct so we can predict the formulas of the new compounds. The oxidation state of copper is obvious, all the alkali earth metals take on a +2 oxidation state, and all the halogens take on a −1 oxidation state, but we'd have to have memorized that Al always takes on a +3 oxidation state, and that the sulfate ion, SO_4^{2-}, has a −2 charge.

A trick for remembering the oxidation states of Ag, Zn, and Al: they are connected diagonally on the periodic table, and as we progress from Ag to Zn to Al, the oxidation states are +1, +2, and +3. Many transition metals take on more than one oxidation state, but Ag and Zn are two important exceptions. So much so that the parentheses that indicate the oxidation states of transition metals in ionic compounds are not used for compounds with Ag and Zn.

174. (D) Ideal solutions are like ideal gases, they don't exist. However, they are useful imaginary models for predicting the properties of real solutions (and gases). An *ideal solution* is one in which each of the particles in solution is subject to the same forces it would be in its pure state. In other words, the different molecules present in an ideal solution have no greater or lesser attractions for the other molecules in the solution as they do for their own kind. There are some assumptions about ideal solutions we should be familiar with: (1) The volume of the solution is purely the sum of the volumes combined. There is no expansion or contraction upon mixing. (2) The heat of solution is zero. It's neither exothermic nor endothermic. (3) All the components of the solution obey Raoult's law (the vapor pressure of each component of the solution is proportional to the mole fraction of that component).

The vapor pressure above an ideal solution is ideal, too—the total pressure is the sum of the partial pressures of the constituent gases (they obey Dalton's law of partial pressures).

The pairs of liquids in choices (C) and (E) are not miscible so we can eliminate them immediately. Choice (A) is incorrect because HCl has a fairly exothermic dissolution (violates assumption No. 2) and it also dissociates into ions, which form ion-molecule attractions with water instead of hydrogen bonds (violates assumption No. 1). Finally, HCl is a gas under standard conditions and this is fairly volatile when in a solution. Choice (B) is tempting because it is an alcohol that can hydrogen bond with water, however the CH_3CH_2– group is nonpolar and would have a greater attraction for other CH_3CH_2OH molecules than it would for water. It's a tough choice, but overall, the pairs of liquids in choice (D) are more similar than the pairs of liquids in choice (B).

175. (D) The most direct and efficient method to determine the *molarity* of the solution is to measure the mass and volume of a sample of the solution (or, the whole solution). We need two pieces of information to get the unit of molarity (M), mol and L. If a solution contains 10 percent hexane by mass, let's assume we have a 100 g sample. That's 10 g, or 0.12 mole, of hexane. The volume of the sample will then allow us to calculate molarity. To make sure we answer a question like this correctly, we can try imagining the situation. We'll find with this problem that we *must know the mass of the sample to calculate the number of moles of solute.* Once we've brought in the assumption of a 100-gram sample, we've admitted that a mass is needed.

176. (A) The trick in this question is in the units of glucose—*milligrams*. The obvious mistake is assuming we have 2 moles of glucose: $^{360}/_{180}$, but we really have $^{2}/_{1,000}$ mole glucose, or 0.002 mole, dissolved in 200 mL of water, or 0.2 L. The final molarity of the solution is $^{0.002}/_{0.2} = 0.01$ M solution. Because glucose is very soluble and the solution is homogenous, any sized sample will have the same concentration.

177. (C) The solution with the highest boiling point will be the one with the greatest number of particles. A simple calculation is to multiply $(m) \times (i)$ (i = the van Hoff factor, the ratio of the number of particles a compound produces when dissolved versus the number of particles of compound added).

	m	i	Total # Moles Particles	Particles
NaCl	0.2	2	0.4	Na^+, Cl^-
$CaCl_2$	0.3	3	0.9	Ca^{2+}, 2 Cl^-
K_3PO_4	0.4	4	****1.6****	3 K^+, 1 PO_4^{3-}
$NaNO_3$	0.5	2	1.0	Na^+, NO_3^-
$C_{12}H_{22}O_{11}$	0.6	1	1.2	$C_{12}H_{22}O_{11}$

178. (C) To solve this problem, we must remember Raoult's law: The vapor pressure of an ideal solution (see **Answer 174** for an explanation of ideal solution) is dependent on the vapor pressure of each component and the mole fraction of the each component. Having 2 moles propylene glycol and 8 moles of water means that out of the 10 moles of total solution, 20 percent is propylene glycol and 80 percent is water. The effect of the solute (propylene glycol) is to lower the vapor pressure of the water because the solute is *nonvolatile*, that is, it doesn't readily vaporize. If the solution is 20 percent nonvolatile solute, then the vapor pressure of the solution will be 20 percent lower than the vapor pressure of pure water. 20 percent of 20 mmHg is 4 $\therefore$ $20 - 4 = 16$ mmHg.

179. (C) The precipitate that formed after the addition of HCl indicates that the solution contained an ion that was insoluble as a chloride. Ag^+, Pb^{2+}, and Hg^{2+} immediately spring to mind. Out of those, only Ag^+ forms a soluble complex ion with NH_3, $Ag(NH_3)^{2+}$. The ion that remained in solution must have been soluble in chloride but insoluble as a sulfate. Out of the ions that are not soluble as a sulfate, Ba^{2+}, Ca^{2+} (not an answer choice), Pb^{2+}, and Ag^+, Ag^+, and Pb^{2+} are not soluble as chlorides and would have already precipitated (as Ag^+ did).

180. (B) All ammonium and nitrate salts are soluble, as are all the group 1 salts. That leaves $BaCO_3$ as the insoluble compound.

Chapter 5: Chemical Reactions

181. (E) H_2O is 18 g mol^{-1}, so 180 g water is 10 moles of water, *one order of magnitude greater* than 1 mole (6.02×10^{23} molecules) $\therefore$ 6.02×10^{24} molecules.

182. (B) Carbon dioxide (CO_2) has a molar mass of 44 g mol^{-1}, so 4.4 g = 0.1 mole (6.02×10^{22} CO_2 molecules). Each CO_2 molecule contains two oxygen atoms, so there are 0.2 mole oxygen atoms or $(0.2) \times (6.02 \times 10^{23}) = \mathbf{1.2 \times 10^{23}}$ atoms of oxygen.

183. (A) Ribose has a molar mass of 150 g mol^{-1}, so 1.5 g = 0.01 mole ribose (6.02×10^{23} ribose molecules). Each ribose has 10 hydrogen atoms, so there are $0.01 \times 10 = 0.1$ mole hydrogen atoms (6.02×10^{22} hydrogen atoms), *one order of magnitude less* than 1 mole (6.02×10^{23} molecules).

184. (E) The compound is $Ti(CO)_6$, titanium hexacarbonyl.

Assume a 100-g sample to convert percent to number of grams. Then convert grams to moles and find the simplest, whole number mole ratio between them.

22.2 g Ti @ 48 g mol^{-1} = ~0.5 moles Ti ÷ 0.5 = 1

33.3 g C @ 12 g mol^{-1} = ~3 moles ÷ 0.5 = 6

44.4 g O @ 16 g mol^{-1} = ~3 moles ÷ 0.5 = 6

Remember to use the mass of atomic oxygen (16 g mol^{-1}) when calculating its percent in a compound.

185. (A) Assume a 100 g sample, convert to moles and find the simplest, whole number mole ratio between them.

92 g C × (12 g mol^{-1}) = ~8 moles

8 g H × (1 g mol^{-1}) = ~8 moles

There is a 1:1 ratio of carbon-to-hydrogen atoms, so CH is the empirical formula.

186. (C) The molecular formula of a compound is a whole number multiple of its empirical formula. If we calculate the molar mass of each of the answer choices, CHO = 29 g mol^{-1}, C_2H_3O = 43 g mol^{-1}, **CH_2O = 30 g mol^{-1}**, and CH_2O_2 = 46 g mol^{-1}. CH_2O, at 30 g mol^{-1}, is the only choice that 150 g mol^{-1} can divide into without a remainder (5 *x*, it is also the empirical formula for the monosaccharides, a class of carbohydrates). Choice (E) is not an empirical formula, it is the molecular formula of ribose ($CH_2O \times 5 = C_5H_{10}O_5$).

187. (E) We're already given the number of mole of each element, so we only need to find the simplest, whole number ratio between them.

0.2 mole Pd ÷ 0.2 = 1

0.8 mole C ÷ 0.2 = 4

1.2 moles H ÷ 0.2 = 6

0.8 mole O ÷ 0.2 = 4 ∴ $PdC_4H_6O_4$, but we can factor 2 from the subscripts on the C, H, and O (the nonmetals which are forming a complex around the Pd cation) to $Pd(C_2H_3O_2)_2$, palladium acetate. The formula $Pd(C_2H_3O_2)_2$ is much more structurally informative than $PdC_4H_6O_4$ because it highlights the fact that there are two acetate ions attached to each Pd^{2+} ion.

188. (D) Assume a 100-g sample, convert to mole and find the simplest, whole number mole ratio between them.

38 g F @ 19 g mol^{-1} = 2 moles ÷ 0.5 = 4

62 g Xe @ 131 g mol^{-1} = 0.5 mole ÷ 0.5 = 1 ∴ XeF_4

189. (B) If a hydrocarbon is 75 percent carbon, it must be 25 percent hydrogen (100% – 75%). Assume a 100-g sample, convert to mole and find the simplest, whole number molar ratio between them.

75 g C @ 12 g mol^{-1} = 6.25 moles ÷ 6.25 = 1

25 g H @ 1 g mol^{-1} = 25 moles ÷ 6.25 = 4 ∴ CH_4

190. (B) The excess H_2 tells us that this is not a limiting reactant problem. The molar mass of Cu_2O is 143 g mol^{-1} but that piece of information is not needed to solve the problem. It is a red herring. What we want to know is that 0.05 mol Cu_2O has 0.1 mol Cu, since there are 2 moles Cu atoms per mole of Cu_2O. The molar mass of Cu is 63.5 g mol^{-1}, so 0.1 mole weighs 6.35 g.

191. (D) A coordination complex consists of an atom or ion surrounded by an array of ions or molecules (called *ligands*). The central atom or ion is typically a metal (most often a transition metal) and the complex it creates is often charged, which is a clue for identifying the complex. In choice (D), Pt is the central metal and the chloride ions are its ligands.

192. (E) Chlorine has an oxidation state of 0 in Cl_2, –1 in Cl^- and +5 in ClO^-.

193. (A) An acid and a base are the reactants of a neutralization reaction. A salt and water are the products.

194. (B) A precipitation reaction produces a solid (if the state of matter of the products is not specified, look for an insoluble compound).

195. (C) Combustion reactions consume O_2 as a reactant and produce oxides.

196. (A) An acidic salt is formed when a strong acid reacts with a weak base. Weak bases form good conjugate acids, whereas strong acids form weak conjugate bases. For example, if Cl^- was able to pick up H^+ in solution, then HCl wouldn't be a strong acid because it wouldn't fully dissociate in solution. The Cl^- would pick up H^+ ions and an equilibrium between HCl and $H^+ + Cl^-$ would be reached. The equilibrium of a strong acid/base ionization isn't a true equilibrium. The reaction goes to completion. The concentrations of products and reactants don't change, but only because the reverse reaction doesn't occur, not because the forward and reverse reactions occur at equal rates. In fact, the concentration of reactant in strong acid or base dissociations is practically zero.

197. (E) Solid aluminum is not oxidized during this reaction. (See **Answer 11** for an *except* question strategy.)

Important Note: Questions like this will typically not appear in the multiple choice section of the exam. The free response section, however, has a mandatory "Reactions" section that requires test takers to write out a balanced equation and answer a question about it.

198. (B) Sodium carbonate is not produced by the heating of sodium bicarbonate. (See important note following **Answer 197**.)

199. (A) The compound P_4O_3 does not form under the conditions in question. It is not a naturally occurring oxide of phosphorus. (See important note following **Answer 197**.)

200. (A) Choices (B) through (D) contain the spectator ions of the total ionic reaction. Choice (E) involves the formation of HI, a strong acid, which is not produced in this reaction. (See important note following **Answer 197**.)

201. (E) $HCl_{(aq)}$ and $KOH_{(aq)}$ = The exothermic neutralization would increase the temperature of the reaction vessel.

$CaCO_{3(aq)}$ and $HF_{(aq)}$ = Bubbling from the carbon dioxide gas would be observed.

$Mg_{(s)}$ and $HI_{(aq)}$ = Bubbling of the hydrogen gas would be observed.

$Pb(NO_3)_{2(aq)}$ and $NaCl_{(aq)}$ form an insoluble precipitate ($PbCl_2$).

The mixing of $NH_4NO_{3(aq)} + HCl_{(aq)}$ solutions is not very exciting. They are both very soluble in water and a double replacement switch produces no insoluble compounds. Nitrate (NO_3^-) and chloride (Cl^-) are poor conjugate bases and will not pick up protons in solution to form HNO_3 or HCl, two strong acids (see **Answer 196** for an explanation of why strong acids form the weakest conjugate bases). Ammonium chloride is a completely soluble salt. (See **Answer 11** for an *except* question strategy.)

202. (C) A combustion reaction is a vigorous, exothermic, self-sustaining reaction in which the substances combine to give off heat and light. The combustion reactions we are most familiar with use an oxidizing agent (like O_2). Choice (E) is an oxidation, but the oxidizing agent in not represented, it is an electrochemical half-reaction.

203. (B) It's the only reaction that produces an ion, CdI_4^{2-}.

204. (E) Remember OIL RIG: **O**xidation **I**s **L**oss of electrons, **R**eduction **I**s **G**ain of electrons.

205. (A) Carbonates form CO_2 gas when acidified according to the equilibrium equation:

$$H^+_{(aq)} + CO_3^{2-}{}_{(aq)} \Leftrightarrow H^+_{(aq)} + HCO_3^-{}_{(aq)} \Leftrightarrow H_2CO_3 \Leftrightarrow CO_{2(g)} + H_2O_{(l)}$$

206. (D) When balancing the combustion of carbon-containing compounds, balance **c**arbon atoms first, **h**ydrogen atoms second, and **o**xygen atoms last (remember Mrs. **Cho**, she'll help you balance combustion reactions).

$$\underline{2}\ C_6H_{14} + \underline{19}\ O_2 \rightarrow \underline{12}\ CO_2 + \underline{14}\ H_2O$$

207. (C) When balancing the combustion of carbon containing compounds, balance **c**arbon atoms first, **h**ydrogen atoms second, and **o**xygen atoms last (remember Mrs. **Cho**, she'll help you balance combustion reactions).

$$\underline{1}\ CH_3CH_2OCH_2CH_{3(g)} + \underline{6}\ O_{2(g)} \rightarrow \underline{4}\ CO_{2(g)} + \underline{5}\ H_2O_{(g)}$$

208. (C) This reaction can be quickly balanced by noticing that the H_2O on the left side of the arrow has the hydrogen and oxygen atoms that will be distributed of NH_3 and OH^-. On the right side, we can see that each water molecule breaks up into a hydrogen ion and a hydroxide ion, so we need 3 water molecules because the N from Li_3N will need 3 hydrogen ions to form ammonia.

$$\underline{1}\ Li_3N_{(s)} + \underline{3}\ H_2O_{(l)} \rightarrow \underline{3}\ Li^+_{(aq)} + \underline{3}\ OH^-_{(aq)} + \underline{1}\ NH_{3(g)}$$

209. (B) To balance redox reactions, it is typically easiest to use oxidation states. We know that $Cr^{3+}_{(aq)}$ needs 3 electrons to be reduced to $Cr_{(s)}$. The confusing part may be with the chlorine ions. Each $Cl^-_{(aq)}$ needs to lose 1 electron but we need them to combine in pairs to produce $Cl_{2(g)}$, so we should immediately put a 2 near $Cl^-_{(aq)}$ to remind us later that we accounted for chlorine's electrons in pairs. We know we need 3 electrons to reduce chromium and 2 electrons to oxidize the chlorines. The least common multiple of 3 and 2 is 6 electrons, and we see that if we lost 6 electrons from 6 chlorine ions (or 3 pairs of chlorines), we'd be able to reduce 2 chromium ions.

$$\underline{2}\ Cr^{3+}_{(aq)} + \underline{6}\ Cl^-_{(aq)} \rightarrow \underline{2}\ Cr_{(s)} + \underline{3}\ Cl_{2(g)}$$

210. (D) When balancing an equation that contains a polyatomic ion, keep the ion together when possible. For example, the first thing we might balance is the magnesium. Once the 3 is in front of $Mg(H_2PO_4)_{2(s)}$, we are left with 6 $H_2PO_4^-$ ions. It's easy to see that the phosphate from magnesium phosphate picked up a hydrogen ion from phosphoric acid (H_3PO_4), so we can work backward. There is a 2:1 ratio of phosphates between $Mg_3(PO_4)_{2(s)}$ and $H_3PO_{4(l)}$, so the 6 we have in the products should be allocated accordingly. Six phosphates can be allocated to 4 phosphoric acid molecules and the other two come from one magnesium phosphate.

$$\underline{1}\ Mg_3(PO_4)_{2(s)} + \underline{4}\ H_3PO_{4(l)} \rightarrow \underline{3}\ Mg(H_2PO_4)_{2(s)}$$

211. (D) *Important note: A type of balancing that appears on the AP Chemistry exam is a redox reaction occurring in an acidic or basic solution. If a question like this appears in the multiple choice section, just balance as you normally would. If the number of each atom on the reactant side of the equation is identical to those in the products, nothing else is needed. However, if it doesn't work, or the question appears in the free-response section (probably associated with an electrochemistry problem) and asks our work be shown, we should be familiar with the procedure.*

To balance redox reactions that occur in acidic and neutral conditions: (1) Balance the atoms *other than oxygen and hydrogen.* (2) Balance oxygen atoms by adding water molecules. (3) Balance hydrogens by adding hydrogen ions. (4) Balance the charges by adding electrons.

$$\underline{3}\ CH_3CH_2OH + \underline{2}\ Cr_2O_7^{2-} + \underline{16}\ H^+ \rightarrow \underline{3}\ CH_3COOH + \underline{2}\ Cr^{3+} + \underline{11}\ H_2O$$

212. (C) See important note in **Answer 211. To balance redox reactions that occur in alkaline conditions:** (1) Balance atoms *other than oxygen and hydrogen.* (2) Balance oxygen and hydrogen atoms *at the same time.* To balance oxygen, we can add hydroxide ions or water, but they both contain oxygen and hydrogen. However, there is a 1:1 hydrogen-to-oxygen ratio in OH^- and the 2:1 hydrogen-to-oxygen ratio in H_2O. To add a hydrogen atom, add a water molecule, the only way to be "one up" on hydrogen in the hydrogen-to-oxygen ratio.

$\underline{1}\ MnO_2 + \underline{2}\ OH^- + \underline{1}\ O_2 \rightarrow \underline{1}\ MnO_4^{2-} + \underline{1}\ H_2O$

213. (C) When given the amounts of both reactants in a problem, check for limiting reactants. A simple way to do this is to take the number of mole of each reactant and divide it by its stoichiometric coefficient in the balanced equation. In this case, 0.4 mole $CS_2 \div 1 = 0.4$, and 1.20 moles $O_2 \div 3 = 0.4$. *Because their quotients are equal, there is no limiting reactant.* The two reactants are available in the correct ratio. If there is no limiting reactant, use the reactant that has the easiest numbers to crunch. In this case, it is 0.4 mole CS_2. One mole CS_2 would produce 3 moles of products (1 mole of CO_2 and 2 moles of SO_2) so 0.4 mole CS_2 will produce 1.2 moles of products.

214. (D) Both the products are gases and the total number of them produced is 3, which also happens to be the number of moles of O_2 required to produce them. So the number of moles of gas produced will be the same as the number of moles of O_2 that reacted.

215. (A) There are 3 moles of gas produced for every mole CS_2 reacted. If 33.6 L of gas formed at STP, then 1.5 moles of gas is produced ($33.6\text{ L} \times (^{1\text{ mol gas}}/_{22.4\text{ L}})$). The math is easy: 1.5 is half of 3, so 0.5 mole of CS_2 reacted (or, 1.5 moles products $\times$ $(^{1\text{ mol }CS_2}/_{3\text{ mol products}})$).

216. (B) One hundred mL (0.1 L) of 0.6 M HCl contains 0.06 mole H^+ ions ($0.1\text{ L} \times (^{0.6\text{ mol HCl}}/_{1\text{ L}})$) which will get reduced to 0.03 mole H_2 gas (Zn is above hydrogen in the activity series and will lose electrons to H^+ ions in an acidic solution). Because the reaction occurs at STP, we don't need to use a gas law, only the conversion factor of 1 mole gas = 22.4 L $\therefore$ $0.03\text{ mol} \times {}^{22.4\text{ L}}/_{\text{mol}} = 0.672\text{ L}$, or 672 mL.

For **Questions 217–219:**

(A) 4 SO_3 only
(B) 2 SO_2 and 2 SO_3
(C) 3 SO_2, 1 O_2 and 2 SO_3
(D) 3 SO_2 and 2 O_2
(E) 4 O_2 and 5 SO_3

217. (D) The reaction of 3 moles of SO_2 with excess oxygen would produce 3 moles of SO_3. Since 1 mole of O_2 is needed for the reaction, and the question asks for 1 mole O_2 in excess, we need to start with 2 moles of O_2. Choice (D) is the only one that contains only reactants (see list above for the contents of each box), so we could have easily chosen the correct answer without even doing the stoichiometry.

218. (B) A simple way to check for a limiting reactant is to take the number of moles of each reactant provided in the question and divide it by its stoichiometric coefficient in the balanced equation: 4 mol $SO_2 \div 2 = 2$ and 1 mol $O_2 \div 1 = 1$ $\therefore$ **O_2 is the limiting reactant**, so we won't get 4 moles of SO_3, we'll get as much SO_3 as what 1 mole of O_2 will produce with excess SO_2 (2 moles). We started with 4 moles of SO_2 and 2 moles were consumed to produce 2 moles of SO_3, leaving 2 moles in excess. (See list above for the contents of each box.)

219. (E) A simple way to check for a limiting reactant is to take the number of moles of each reactant provided in the question and divide it by its stoichiometric coefficient in the balanced equation: 5 mol $SO_2 \div 2 = 2.5$ and 6.5 mol $O_2 \div 1 = 6.5$, so **SO_2 is our limiting reactant.** With 5 moles of SO_2, we can produce 5 moles of SO_3 and will consume 2.5 moles of O_2 in the process. That leaves us with 4 moles of O_2 in excess. (See list above for the contents of each box).

220. (B) When we are presented with the quantities of two or more reactants in a problem, we check for a limiting reactant. A simple way to check for a limiting reactant is to take the number of moles of each reactant provided in the question and divide it by its stoichiometric coefficient in the balanced equation: 0.1 mol $NH_3 \div 4 = 0.025$, 0.1 mol $O_2 \div 5 = 0.02$ $\therefore$ **O_2 is our limiting reactant.** The maximum amount of NO that can be produced with 0.1 mole of O_2 is 0.08 mole of NO (0.1 mol $O_2 \times {}^{4 \text{ mol NO}}/_{5 \text{ mol } O_2}$), or 2.4 g NO.

221. (E) The reaction of NH_3 and O_2 proceeds according to the equation:

$$4\ NH_3 + 5\ O_2 \rightarrow 4\ NO + 6\ H_2O$$

The 4:5 molar ratio given in the question are the exact coefficients of the reactants in the balanced equation, which makes the problem fairly simple. Because the problem states that the reactants reacted 100 percent, only a ratio that reflects the same relationship as the stoichiometry in the balanced equation could be given.

The products, NO and H_2O, will be present in a 4:6 ratio as given in the balanced equation. Remember, however, that these are mole ratios, *not* mass ratios. The molar mass of NO = 30 g mol^{-1} and the molar mass of H_2O = 18 g mol^{-1}. 4 moles NO = 120 g and 6 moles H_2O = 108 g. 120 g + 108 g = 228 g, the exact yield of the reaction. If the numbers given in a problem don't align perfectly with the balanced equation, use the mass stoichiometry instead. For example:

$$\mathbf{4\ NH_3 + 5\ O_2 \rightarrow 4\ NO + 6\ H_2O}$$

$$(4 \text{ moles } NH_3)\ (17 \text{ g mol}^{-1}) + (5 \text{ moles } O_2)\ (32 \text{ g mol}^{-1}) \rightarrow$$

$$(4 \text{ moles NO})\ (30 \text{ g mol}^{-1}) + (6 \text{ moles } H_2O)\ (18 \text{ g mol}^{-1})$$

$$68 \text{ g } NH_3 + 160 \text{ g } O_2 \rightarrow 120 \text{ g NO} + 108 \text{ g } H_2O$$

Now we can determine the percent mass of each product:

$${}^{120 \text{ g NO}}/_{228 \text{ g total}} = \sim 53\% \text{ NO} \therefore 47\%\ H_2O$$

You *must* use the 228 g from the addition of 120 g NO + 108 g H_2 from the equation to get the correct percentages. If we were given another mass of product, let's say 72 g, we would take the 53 percent we arrived at by the method above and then apply it to the mass given in the problem: 53 percent of 72 g = ~38 g NO $\therefore$ 34 g H_2O.

222. (C) We use $M_1V_1 = M_2V_2 \therefore (0.375)\ (400) = (x)\ (500) \therefore x = 0.3$ M. *Add volumes* when combining solutions.

223. (E) Some questions ask us to *set up a problem* as opposed to actually calculating an answer. In this example, start by setting up the expression to convert 1.0 L of NO into moles of NO ($1.0\ L \times {}^{1\ mol}/_{22.4\ L}$). Since we need 22.4 in the denominator, we can immediately eliminate choices (A) and (B). Next, arrange an expression that converts mol of NO produced into mol O_2 consumed using the stoichiometric coefficients from the balanced equation (moles NO produced $\times {}^{5\ mol\ O_2\ consumed}/_{4\ mol\ NO\ produced}$ = moles O_2 consumed). We need the fraction ${}^5/_4$, which leaves (E) as our only choice.

224. (B) When presented with the quantities of two or more reactants in a problem, check for a limiting reactant. A simple way to check for a limiting reactant is to take the number of moles of each reactant provided in the question and divide it by its stoichiometric coefficient in the balanced equation. 2.5 mol $NH_3 \div 4 > 2.5$ mol $O_2 \div 5$, therefore **O_2 is our limiting reactant** and NH_3 would remain in excess.

$$2.5 \text{ moles } O_2 \text{ available} \times {}^{4\ mol\ NH_3\ consumed}/_{5\ mol\ O_2\ consumed} = 2 \text{ moles } NH_3 \text{ consumed}$$

We started with 2.5 moles NH_3 ∴ 0.5 mole remains.

225. (A) When presented with the quantities of two or more reactants in a problem, check for a limiting reactant. A simple way to check for a limiting reactant is to take the number of moles of each reactant provided in the question and divide it by its stoichiometric coefficient in the balanced equation: 6 moles $KO_2 \div 4 < 9$ moles $CO_2 \div 2$ ∴ **KO_2 is our limiting reactant**.

$$6 \text{ mol } KO_2 \times {}^{3\ mol\ O_2}/_{4\ mol\ KO_2} = 4.5 \text{ moles } O_2 \text{ produced}$$

226. (A) When carrying out stoichiometry with gases under the same conditions (temperature and pressure), we can treat volumes like moles (at the same temperature and pressure, equal volumes of any gases have the same number of particles).

$$6\ L\ O_2 \times {}^{2\ CO_2}/_{3\ O_2} = 4\ L\ CO_2$$

227. (A) Because we are given answers in L, it is easiest for us to convert CO_2 to L first, and then use the gas stoichiometry shortcut described in **Answer 226** (treat volumes like moles).

$$2.9\ g\ CO_2 \times {}^{1\ mol}/_{44\ g} \times {}^{22.4\ L}/_{1\ mol} = \sim 1.5\ L\ CO_2$$

$$1.5\ L\ CO_2 \times {}^{3\ O_2}/_{2\ CO_2} = 2.25\ L\ O_2$$

228. (A) The net ionic equation for just about every neutralization reaction is $H^+ + OH^- \rightarrow H_2O$. Na^+ and NO_3^- ions are (always) spectator ions. They remain in solution (in other words, they don't react) for the entire reaction.

229. (B) An *addition reaction* is mainly limited to alkenes and alkynes. A typical addition reaction on the AP Chemistry exam will add a halogen to an alkene. Reaction (A) is a substitution reaction. A *substitution reaction* occurs when a functional group (or a hydrogen atom, as in this reaction) is replaced by another group (or element, often a halogen). These reactions mainly involve alkanes. Reaction (D) is *saponification*, the process that produces soap from fat and a strong base (typically NaOH, also called lye). The result is a soap of the carboxylate (in this case, it is a fatty acid, a long hydrocarbon chain with a carboxyl group at the end). Reaction (E) demonstrates photosynthesis.

230. (E) The reaction between carbon dioxide and water is worth memorizing.

$$CO_{2(g)} + H_2O_{(l)} \Leftrightarrow H_2CO_3 \Leftrightarrow H^+_{(aq)} + HCO_3^-{}_{(aq)} \Leftrightarrow H^+_{(aq)} + CO_3^{2-}{}_{(aq)}$$

Chapter 6: Thermodynamics

231. (B) The definition of lattice energy (kJ mol^{-1}). (See **Answer 52** for a description of the factors that determine lattice energy.)

232. (C) The definition of (Gibb's) free energy (ΔG, kJ mol^{-1}). (See **Answer 239** for an explanation of Gibb's free energy.)

233. (A) A definition of activation energy, E_a. (See **Question 234** for another definition of E_a and **Answers 235, 241, and 251** for descriptions of the different aspects of activation energy.)

234. (A) Another definition of activation energy, E_a. (See **Question 233** for another definition of E_a and **Answers 235, 241, and 251** for descriptions of the different aspects of activation energy.)

235. (A) Yet a third way to consider the *activation energy, E_a*, of a reaction (see **Question 233 and 234**) is that it reflects the sensitivity of the reaction rate to temperature changes. The *Arrhenius equation* quantifies the relationship between activation energy and the rate at which a reaction proceeds: $E_a = RT \ln (^k/_A)$, where R is the universal gas constant, T is the temperature, k is the rate constant, and A is the *frequency factor* for the reaction, a constant for a given reaction that represents an empirical relationship between temperature and the rate constant and has the units s^{-1}. (See **Answer 241** for an explanation of how the Arrhenius equation illuminates the relationship between E_a and K_{eq}.)

236. (E) Substances undergoing phase changes don't experience a change in temperature, an indication of the average kinetic energy of the particles. Instead, the potential energy of the particles changes as they change positions relative to one another.

237. (D) The formula for kinetic energy is ½ mv^2. Graham's law of effusion allows us to calculate the relative speeds of two different gases at the same temperature and is derived from the formula for kinetic energy, which can be rearranged to solve for the average velocity, v, of the particles. Because they are at the same temperature, the value for kinetic energy (which is proportional to but not equal to the absolute temperature) is the same for gases 1 and 2:

$$\tfrac{1}{2} m_1 v_1^2 = \tfrac{1}{2} m_2 v_2^2$$

Solving for v_1/v_2: $^{v_1}/_{v_2} = \sqrt{(^{m_2}/_{m_1})}$

The velocity is inversely proportional to the square root of the molar mass. (See **Answer 123** for an in-depth explanation of effusion.)

238. (B) The definition of zero entropy is a perfect, pure crystalline solid at 0 K (and the third law of thermodynamics).

239. (E) *Gibb's free energy* is a thermodynamic state function given by the formula $\Delta G = \Delta H - T\Delta S$. It is a measure of how much useful (but nonmechanical) work can be done by a system (at a constant pressure and temperature). Systems in disequilibrium, that is, not at equilibrium, have the ability to do work as they move toward equilibrium. Once at equilibrium, no more work can be done. A useful form of the Gibb's free energy equation, $\Delta G° = -RT \ln K_{eq}$, allows the standard-state free energy change of a reaction to be calculated if the K_{eq} is known. More importantly, it shows that *the equilibrium established for a reaction is a function of the free energy change* (for reactions in solution).

240. (A) The standard enthalpy of formation ($\Delta H°_f$) is the enthalpy change that accompanies the formation of one mole of a compound (or an element in its nonstandard state) from its standard-state elements. For example, $\Delta H°_f C_{(g)} = 0$ kJ mol^{-1}, $\Delta H°_f O_{2(g)} = 0$ kJ mol^{-1} but the $\Delta H°_f CO_{2(g)} = -393.5$ kJ mol^{-1}, which tells us that the formations of one mole of carbon dioxide gas from 1 mole of $C_{(g)}$ and 1 mole of $O_{2(g)}$ is exothermic and produces 393.5 kJ of heat.

241. (D) *The Arrhenius equation* quantifies the relationship between the activation energy, E_a, and the rate at which the reaction proceeds: $E_a = -RT \ln (^k/_A)$ where R is the universal gas constant, T is the temperature (in K), k is the rate constant (from the rate law) for the reaction, and A is the frequency factor for a particular reaction. It is a quantity related to the frequency of collisions between particles that are correctly oriented to produce an effective collision. Any reaction with a positive activation energy (the vast majority) will experience an increased rate of reaction with increasing temperature. (See **Answer 235** for an explanation of how E_a is measured, **Answer 251** for a practical description of activation energy, and **Answers 277 and 278** for further explanations of temperature, activation energy, and reaction rate.)

242. (D) An exothermic reaction that increases entropy will be spontaneous (exergonic) at all temperatures. In general, nature prefers moving toward a lower energy and more entropic state. (See table below.)

ΔG	ΔH	ΔS	T
– (spontaneous)	– (exothermic)	+ (⇑ entropy)	all
+ (nonspontaneous)	+ (endothermic)	– (⇓ entropy)	all
– (spontaneous)	– (exothermic)	– (⇓ entropy)	low only
+ (nonspontaneous)	+ (endothermic)	+ (⇑ entropy)	high only

243. (A) Melting is an endothermic reaction (heat must be added) that increases entropy.

SOLID	⇒	**LIQUID**	⇒	**GAS**
Least entropic	(add heat)		(add heat)	*Most entropic*

244. (B) Both deposition and condensation are exothermic processes that decrease entropy.

GAS	$\Rightarrow$	LIQUID	$\Rightarrow$	SOLID
Most entropic	(remove heat)		(remove heat)	*Least entropic*

245. (D) Wood is mostly made of lignin and cellulose. Lignin is a complex chemical compound (its formula is $C_9H_{10}O_2$, $C_{10}H_{12}O_3$, $C_{11}H_{14}O_4$) and cellulose is a large, linear polymer of glucose $(C_6H_{12}O_6)_n$. Their combustion with O_2 produces CO_2 and H_2O as follows (for one glucose):

$$C_6H_{12}O_{6(s)} + 6\ O_{2(g)} \rightarrow 6\ CO_{2(g)} + 6\ H_2O_{(g)}$$

The combustion of wood (or petroleum products) is highly exothermic (which is why we use it for heat, light, and work) and produces six more moles of gas than it consumes, which results in increased entropy. In addition, the high heat produced is also highly entropic. Remember, *the definition of zero entropy is a perfect, pure crystalline solid at 0 K* (it's also the third law of thermodynamics). Any deviation from 0 K and/or a pure, perfect crystalline solid indicates that entropy is increasing. An increase in temperature typically increases the entropy of the system. (See table below **Answer 242**.)

246. (C) A reaction that is both endothermic and results in decreased entropy will never be spontaneous at any temperature (like the formation of wood from $CO_{2(g)}$ and $H_2O_{(g)}$. Plants do it all the time, but it takes a lot of energy (from the sun) to do it. In general, nature moves toward a state that is of lower energy and higher entropy. When an entropy reduction occurs in one process (organizing the chemistry lab), it does so by creating *more* entropy in the universe. (The heat created by our energy metabolism, for example. Our mitochondria are only ~40 percent efficient at converting food energy into cellular energy, the rest is lost as heat, the most entropic form of energy.) (See table below **Answer 242**.)

247. (E) The mixing of two gases at a constant temperature (an adiabatic process is one that occurs without heat transfer) must have a $\Delta H = 0$. If no reaction occurs and the gases experience no forces of attraction or repulsion, then the entropy of the system must increase ($\Delta S > 0$). The entropy of a system is proportional to the number of states a system *can* assume versus the number of states that are "right." For example, imagine a strange deck of cards in which each card has a number from 1 through 52 and, all 52 are in numerical order. For simplicity, we will assume this is the *only* ordered state of the deck. If the cards are shuffled in any way, they are out of order, or disordered. It is easy to see that there are there are $(52 - 1)!$ ways the cards are considered disordered and only *one* way they are considered ordered. If there were only two cards in the deck, there would be $(2-1)!$, or one way the cards could be disordered and one way they could be *in order*, so the more (different) components that make up a system, the more potential for disorder and entropy in the system.

248. (C) To calculate the ΔH_{rxn} from $\Delta H°_f$ values, remember a simple formula:

$$\mathbf{\Delta H_{rxn} = (\Delta H°_f PRODUCTS) - (\Delta H°_f REACTANTS)}$$

Add the ΔH°_f of the products and reactants separately, then subtract the value of the sum of the reactants from the sum of the products. *Be careful with signs and make sure to multiply the ΔH°_f of each substance by its stoichiometric coefficient.*

$$\Delta H_{rxn} = (\Delta H^\circ_f \text{PRODUCTS}) - (\Delta H^\circ_f \text{REACTANTS})$$

$$\Delta H_{rxn} = (-635.5 + -393.5) - (-1{,}207.1)$$

$$\Delta H_{rxn} = 178 = \sim 180 \text{ kJ mol}^{-1}$$

249. (C) This is a Hess law problem. Since enthalpy is a state function (it only depends on final and initial states, a D sign is usually a clue), it doesn't matter how we get there, only where we start and where we end up. (*H, G, and S* as well as *T, V, and P* are all state functions, too.)

If we keep the conversion of graphite and oxygen to carbon dioxide, but "flip" the conversion of diamond and oxygen to carbon dioxide (and remember to *reverse the sign of* ΔH), the two equations cancel out oxygen (with a double slash) and carbon dioxide (with a single slash) and their sum leaves us with the conversion of graphite to diamond that we wanted:

$$\mathbf{C}_{(graphite)} + \not{O}_{2(g)} \rightarrow \not{CO}_{2(g)} \Delta H = -393.5 \text{ kJ}$$

$$\not{CO}_{2(g)} \rightarrow \mathbf{C}_{(diamond)} + \not{O}_{2(g)} \Delta H = +395.4 \text{ kJ}$$

$$C_{(graphite)} \rightarrow C_{(diamond)} \Delta H = +1.9 \text{ kJ}$$

250. (D) One way to think of entropy is as a measure of disorder in a system. The third law of thermodynamics states that the entropy of a pure, perfect crystal at 0 K is zero. Any change that takes a system further from that state is typically thought of as increasing entropy. This is mostly true, and is a good guide to predicting whether a particular change will increase or decrease entropy. For example, a mixture typically has more entropy than a pure substance.

Under certain circumstances, however, the entropy changes expected are not what is observed. In this situation, the addition of Ca^{2+} and Cl^- ions *decreases* entropy because the water molecules end up becoming more ordered as they form hydration shells around the ions. The entropy change of a system is the sum of the entropy changes of its components, so even though the $CaCl_2$ became *less* ordered as it dissolved, the magnitude of the entropy decrease of the water molecules was greater than magnitude of the entropy increase of $CaCl_2$. (See **Answer 247** for an explanation of why a mixture is expected to have greater entropy than a pure substance and for a different consideration of entropy.)

251. (D) We can think of the activation energy of a reaction as the energy needed for the formation of the transition state of the reaction. It is considered an energy barrier to the reactions progression. The spark of the lighter produces the heat needed to get the combustion going by supplying the energy needed by the reactants to form the transition state needed to form the products. The combustion of the butane is highly exergonic, so once the reaction starts, the energy it liberates supplies the activation energy for the other butane molecules in the lighter (which are sprayed into the atmosphere where they react with oxygen).

We probably don't go to bed at night worrying that our chemistry textbook will turn into carbon dioxide and water while we sleep. The conditions at our desk don't provide the activation energy needed to start the process. But if we applied the heat of the lighter flames to it (or, if we could get the spark to directly catch one of the pages), we could easily turn our book into carbon dioxide and water (not recommended until *after* the AP Chemistry exam). (See **Answer 235** for an explanation of how E_a is measured, **Answer 241** for the relationship between E_a and K_{eq}, and **Answers 277 and 278** for further explanations of temperature, activation energy, and reaction rate.)

252. (D) (See **Answer 250** for an explanation of entropy and **Answers 243 and 244** for tables of the entropy changes associated with phase changes.)

253. (C) Reaction (C) formed one kind of solid compound from a solid and a gas. (See **Answer 250** for an explanation of entropy, and **Answers 243 and 244** for tables of the entropy changes associated with phase changes.)

254. (E) This is a stoichiometry problem. The tricky part is that the ΔH_{rxn} is given in kJ mol^{-1} CH_6N_2, but the stoichiometric coefficient of CH_6N_2 in the reaction is 2. That means we don't convert 6 moles of $H_2O_{(g)}$ to $H_2O_{(l)}$, we only need to convert 3 (because we're asked per mole of CH_6N_2).

$$H_2O_{(g)} \rightarrow H_2O_{(l)} = -44 \text{ kJ mol}^{-1} \times 3 \text{ mol} = -132 \text{ kJ}$$

The negative sign indicates the condensation of water is exothermic (which we probably already knew), but that means we will actually get *more* energy out of the combustion of CH_6N_2, so we'll want a ΔH value that is *more negative*

$$\therefore -1{,}303 \text{ kJ mol}^{-1} + (-132 \text{ kJ}) = -1{,}435 \text{ kJ mol}^{-1}.$$

255. (A) If a reaction lowers the temperature of the container in which it occurs, the reaction is *endothermic*. It is absorbing heat from its environment; this is why it decreases in temperature (it is losing heat to the reaction). Most dissolutions have a $+\Delta S$, they increase entropy. (See **Answer 250** for an exception.)

256. (A) The units of ΔS are J mol^{-1} K^{-1}, *not* kJ mol^{-1}, like ΔG and ΔH. A spontaneous reaction has a $-\Delta G$. If the reaction is spontaneous *only at low temperatures*, then we know that the reaction is *exothermic* and the entropy change is *positive* (see table below **Answer 242**).

$$\Delta H - T\Delta S < 0$$

$$(-18{,}000 \text{ J mol}^{-1}) - (300)(\Delta S) < 0$$

$$300\ \Delta S < 18{,}000 \ \therefore \Delta S < 60 \text{ J mol}^{-1}$$

257. (D) The double arrow tells us that the reaction is at equilibrium. At equilibrium, $\Delta G = 0$. Using Gibbs free energy equation $\Delta G = \Delta H - T\Delta S$, if $\Delta G = 0$ then $\Delta H = T\Delta S$. For a phase change, the change in entropy is equal to the heat of fusion divided by the melting point temperature (in Kelvin).

258. (C) We live in a fairly stable environment thanks to activation energies. We probably don't go to bed at night worrying that our bed will turn into carbon dioxide and water while we sleep. The conditions in our house don't provide the activation energy needed to start the combustion process. But a bed in the middle of a forest fire would quickly be consumed and become a great mass (and volume) of gases. The higher the E_a of a given reaction, the more thermodynamically stable the reactants and the more energy they need to form the activated complex in their transition to product(s). We literally can sleep at night because of the stability activation energies provide.

259. (E) The short explanation (probably worth memorizing) is that *all adiabatic, reversible process are isentropic* ($\Delta S = 0$, no entropy change). In an adiabatic process there is no heat flow between the system and its surroundings $\therefore \Delta\text{heat} = 0$. If $\Delta S = {}^{\Delta\text{heat}}/_{\Delta T}$ and $\Delta\text{heat} = 0$ $\therefore$ $S = 0$. The entropy of the gas in a cylinder is increased when its temperature is increased and the entropy is decreased when its volume is reduced. When we compress the cylinder of gas without heat exchange, its volume is reduced while its temperature rises, so its entropy is unchanged.

260. (D) *Bond energy is a measure of bond strength.* It is the amount of heat required to break one mole of a particular bond. All bond energies are positive numbers, because *breaking bonds is always endothermic.* We use the formula $\mathbf{H_{rxn}}$ = **(energy of bonds broken) – (energy of bonds formed)** to determine the ΔH_{rxn} using bond energy data. A simple mnemonic device to remember the formula is "B-FOR" (before): bonds broken are accounted for B-FOR bonds FORmed.

For the reaction $2\ H_{2(g)} + O_{2(g)} \rightarrow 2\ H_2O_{(l)}$

The bonds broken are:

2 moles H—H bonds @ 432 kJ mol^{-1} = 864 kJ
1 mole O=O bonds @ 494 kJ mol^{-1} = 439 kJ
Total energy needed to break bonds = 1,358 kJ

The bonds formed are:

4 moles O—H bonds @ 459 kJ mol^{-1} = 1,836 kJ
Total energy released by bond formation = 1,836 kJ

The formula already accounts for the exothermicity of bond formation with the "–" sign, so don't change the signs; leave them positive.

ΔH_{rxn} = (energy of bonds broken) – (energy of bonds formed)

ΔH_{rxn} = (1,358 kJ) – (1,836) = –478 kJ per 2 moles H_2O

$\therefore$ The formation of one mole of H_2O is –239 kJ. Alternatively, we could have performed the calculations for 1 mole H—H bonds and ½ mole O_2 bonds to form 1 mole H_2O.

Break *all* bonds in the reactants and form *all* the bonds in the products unless the reaction mechanism is known with complete certainty. If a bond isn't actually broken during the reaction, its formation will be included in the calculation, so the value of the bond energy of the formed bond will be subtracted from the bond energies of the broken bonds, canceling itself out (the magnitude of the energy change for breaking and forming a particular bond is the same, only the sign changes). For example, to break an H—H bond requires an input of 432 kJ mol^{-1} and forming and H—H bond results in 432 kJ of energy given off. There is no net energy change for breaking and forming the same bond.

261. (D) The ΔH_{rxn} can be calculated by the formula:

$$\mathbf{\Delta H_{rxn} = (\Sigma\Delta H^\circ_{\mathit{f}}\ products) - (\Sigma\Delta H^\circ_{\mathit{f}}\ reactants)}$$

Remember that *the ΔH°_f of elements in their standard states is zero.* The stoichiometric coefficient of each species must be multiplied to the ΔH°_f of that species to correctly calculate the total enthalpy change of the reaction.

$$\Delta H_{rxn} = (\Sigma\Delta H^\circ_f \text{ products}) - (\Sigma\Delta H^\circ_f \text{ reactants})$$

$$\text{Let } \Delta H^\circ_f CO_2 = x \therefore -1{,}367 = (2x + 3(-286)) - (-278 + 0)$$

$$\therefore -1{,}367 = (2x - 858) + 278 \therefore 2x = -787 \therefore x = 393.5 \text{ kJ mol}^{-1}$$

262. (B) We use the formula $\Delta H_{rxn} = (\Sigma\Delta H^\circ_f \text{ products}) - (\Sigma\Delta H^\circ_f \text{ reactants})$

Remember to multiply the ΔH°_f of the compound by the stoichiometric coefficient in the balanced equation to get the total enthalpy change.

$$\Delta H_{rxn} = (9.7 - 2(34)) \therefore \Delta H_{rxn} = -58.3 \text{ kJ}$$

(**Question 261** is similar.)

263. (D) The temperature decrease in the beaker indicates an endothermic reaction (the reaction absorbed heat from the environment), therefore $\Delta H > 0$. The gas liberated indicates that entropy was increased, therefore $\Delta S > 0$.

264. (D) The activation energies (E_a) for forward and reverse reactions are typically different. Activation energy is the energy difference between the reactants and the activated complex. If the forward reaction is exothermic, the E_a = (energy of activated complex) – (energy of reactants). For the reverse, endothermic reaction, the E_a includes the enthalpy change of the reaction. In other words, the E_a of the reverse reaction = ($E_{a\ exothermic\ rxn} + \Delta H_{rxn}$). If the E_a of the forward and reverse reactions are the same, the ΔH of the reaction must be zero.

265. (D) This is a Hess law problem.

The reaction we want is $2\ CO_{(g)} \Leftrightarrow C_{(s)} + CO_{2(g)}$.

If we combine the other two reactions to produce this reaction, we can determine the ΔH_{rxn}. Because enthalpy is a state function (it only depends on final and initial states), it doesn't matter how we get there, only where we start and where we end up.

$$\mathbf{2\ CO_{(g)} \Leftrightarrow C_{(s)} + CO_{2(g)}}$$

$$\mathbf{CO_{(g)}} + \cancel{H_{2(g)}} \Leftrightarrow \mathbf{C_{(s)}} + \cancel{H_2O_{(g)}} \qquad \Delta H^\circ_{298} = -131 \text{ kJ}$$

$$\cancel{H_2O_{(g)}} + \mathbf{CO_{(g)}} \Leftrightarrow \mathbf{CO_{2(g)}} + \cancel{H_{2(g)}} \qquad \Delta H^\circ_{298} = -41 \text{ kJ}$$

We had to reverse both reactions to cancel out intermediates and end up with the correct forward reaction, so we must reverse the signs of $\Delta H°_{298}$ for each reaction and add them $\therefore \Delta H°_{298} = -172$ kJ.

266. (E) The K_p of a gaseous system at equilibrium will *not* change with pressure changes. When the pressure on a gaseous system in equilibrium increases, the equilibrium shifts to favor the side of the reaction with the least number of moles of gases. For a pressure increase to shift the equilibrium, there must be *a different number of moles of gas in the products and the reactants.* In reaction X, there's 1 mole of gas in the reactants and 2 moles of gas in the products, therefore increasing the pressure will *shift the reaction to favor the reactants.* In reaction Y, there is an equal number of moles of gases in the products and reactants, so no shift will occur with a pressure change. In reaction Z, there are 2 moles of gas in the reactants and 1 in the products, so product formation will be favored under increased pressure. However, when the new equilibrium is established, *the value of K_p will be the same.* Even though partial pressures of the individual gases will change, their ratio (as defined by the equilibrium expression) will have the same value for K_p. *The numerical values of K_p, K_{eq}, and K_{sp} for a particular reaction only change with temperature.*

267. (D) *The K_p increases with increasing temperature for endothermic reactions and decreases with increasing temperature for exothermic reactions.* If we imagine heat as a product of an exothermic reaction, adding more heat by increasing temperature pushes the equilibrium to the left, favoring the reactants. The major difference between adding more heat and adding more of a chemical product is that increasing the temperature is accomplished by continually adding more heat, so the reaction remains "pushed" to the left; whereas adding more of a chemical product transiently pushes the equilibrium to the left, but then the prior equilibrium reestablishes itself.

268. (D) Gases are the most entropic state of matter and solids are the least entropic state. Converting a gas to a solid decreases entropy, therefore $S_{final} - S_{initial} < 0$.

269. (E) Thermodynamic data only quantify (or describe) energy and entropy changes in chemical reactions.

Chapter 7: Kinetics

For **Questions 270–273:**

(A) First-order for M, first-order reaction
(B) First order for M and N, second-order reaction
(C) First-order for M and second-order for N, third-order reaction
(D) Second-order for M and first-order for N, third-order reaction
(E) Second-order for N, second-order reaction

270. (E) If changing the concentration of M has no effect on reaction rate, we are looking for a rate law that does not include M.

271. (A) If the total concentration change of 2x doubles the reaction rate, we are looking for a first-order reaction ($[2x]^? = 2$, $\therefore ? = 1$). Only A is a first-order reaction. In this case, the concentration change of N made no difference to the reaction rate, only the concentration change of M.

272. (D) If doubling the concentration of M quadruples the reaction rate, the reaction must be second-order with respect to M ($[2x]^? = 4$, $\therefore ? = 2$). The effect of N was not considered, so we are only concerned with a rate law that contains $[M]^2$.

273. (D) If doubling the concentration of reactants increases the reaction rate by a factor of 8, then the total reaction order must be 3 ($[2x]^? = 8$, $\therefore ? = 3$). If no other information were given, both (C) and (D) are possible rate laws. However, if halving the concentration of N reduces reaction rate by fourfold, then the reactant order for N must be 2 ($[\frac{1}{2} x]^? = \frac{1}{4}$, $\therefore ? = 2$). Alternatively, we can invert the situation to read "doubling the concentration of N increases the reaction rate by fourfold," which is a little easier to consider. We can eliminate (B) because we need a rate law that contains $[N]^2$.

274. (D) This question is asking us about the effect of surface area on reaction rate. As a general rule, increasing the surface area of a solid will increase reaction rate. Breaking up a solid doesn't affect its concentration [choice (C), although concentration isn't a term typically applied to solids]. Choice (A) may seem alluring because it addresses (indirectly) the surface area issue (exposure), but the iron needs to react with *sulfur*, not oxygen, and the oxidation state of iron should be 0 to react with sulfur since it will losing electrons to reduce sulfur. If the iron were partially oxidized, it would be a less effecting reducing agent. (See **Answer 275** for a detailed explanation of effect of surface area on reaction rate.)

275. (B) This question, like **Question 274**, is asking us about the effect of surface area on reaction rate. However, *two* answer choices address surface area. If we had two cubes of wood, one with dimensions $2 \times 2 \times 2$ and other with dimensions $4 \times 4 \times 4$, their surface areas would be 24 and 96, respectively (the units are irrelevant). The surface area of the larger piece of wood is four times that of the smaller piece of wood which makes sense, the cube is basically two times as large and since area has a unit of is a length^2, $(2 \times \text{length})^2 = 4 \times$ the surface area. The volume of the $2 \times 2 \times 2$ cube is 8, whereas the volume of the $4 \times 4 \times 4$ cube is 64. As expected, the volume of the larger cube is eight times that of the smaller cube. The unit of volume is length^3, so $(2 \times \text{length})^3 = 8$.

Increasing surface area increases the rate of a reaction due to the greater exposure *per unit of volume or mass.*

When we revisit our cubes, the $2 \times 2 \times 2$ cube has a surface area to volume ratio of $^{24}/_8 = 3$, but the surface area to volume ratio of the $4 \times 4 \times 4$ cube is $^{96}/_{64} = 1.5$. Notice doubling the size decreases the surface area to volume ratio by one-half.

If the two solids are the same (wood, twigs, sawdust) then the composition (and concentration, which isn't a term typically applied to solids) does not change when we break it into smaller pieces.

276. (D) Increasing the pressure only shifts the equilibrium (and thereby momentarily changes the rate of either the forward or reverse reaction) of a reaction that involves gases, and even then, the number of gaseous moles of reactant and product must differ or the pressure change has no effect. Increasing the temperature at which a reaction occurs will almost always increase the reaction rate, whether the reaction is exothermic or endothermic. (For explanations regarding the effect of surface area on reaction rate, see **Answers 274 and 275**.) (See **Answer 11** for an *except* question strategy.)

277. (D) A very common misconception regarding the effect of temperature on reaction rate is that increased temperatures significantly increases the frequency of collisions. The small increase in collision frequency doesn't actually have much effect on reaction rate.

In order for a chemical reaction to occur, the reactants must collide, but the collision itself will not form a product unless the reactants collide *in the proper orientation* to one another and *with enough kinetic energy*. When reactants collide in the right orientation and with enough energy to overcome the activation energy of the reaction, they produce a product (or a reaction intermediate, at least). This is called an *effective collision*.

Imagine Batman trying to fight off the bad guys under the misconception that more collisions (between him and the bad guy) will result in a higher reaction rate. He calculates that if he uses his fingertips instead of his fists, he can get in five times more collisions than with just his two fists alone. According to his logic, a fight that would have normally taken an hour should take only 12 minutes. He goes off into the night, finds the bad guys, and applies his new logic to their fight. He soon finds that poking is no substitute for punching. And just like collisions between reactants have to occur not only with enough energy but in an orientation that allows the reactants to collide in just the right position relative to each other, a punch delivered to the face or belly is going to be much more effective at stopping the bad guys than a punch in the arm or leg.

278. (C) Temperature increases the reaction rate of almost all chemical reactions, whether they are exothermic of endothermic (there are some cases for which this isn't true, and these reactions have *negative activation energies)*. Higher temperatures result in more *effective collisions* (collisions with enough energy and with the proper orientation or reactants relative to each other) by increasing the average kinetic energy of the particles in the reaction mixture, assuring that a greater number of collisions will occur with an energy that meets or exceeds the *activation energy* of the reaction, the energy barrier that must be overcome in order for the reaction to proceed. (See **Answer 277** for a more in-depth explanation of effective collisions.)

279. (A) This question is very tricky. Choice (A) is *almost* correct—each radioactive species *does* have a particular half-life, but a particular element can have *several* different isotopes, each with its own half-life. (See **Answer 11** for an *except* question strategy.)

280. (C) In comparing Trials 1 and 2, the two trials in which the O_2 concentration does not change, we see that the concentration of NO is increased by 1.6 (not quite 2). The rate increases by about 2.6-fold. If we were allowed to use a calculator (in some sections of the free response section), we could easily plug in ($[1.6x]^?$ = 2.6 fold rate increase). If we encounter a problem like this in the multiple choice section, we do an estimate: If the reactant order for NO were 1, doubling the concentration would double the rate, so a 1.6-fold increase in concentration should increase the rate by 1.6-fold. But the change we see is 2.6-fold. We can probably see that the reaction order isn't 3, but maybe we're unsure of 2. If the reactant order of NO is 2, and the concentration of NO was doubled, we'd expect the rate would increase by fourfold. If the concentration was increased by 1.6, $1.6^2 = 2.56$, or 2.6.

We can apply the same methodology to the reactant order of O_2. The two trials in which *only* the concentration of O_2 changes are Trials 1 and 3. The O_2 concentration increases from 0.035 to 0.045, an increase of about 1.3 times. The rate increases from 0.143 to 0.184. If we recognize that 0.143 is about two-thirds of 0.184, we can determine that the reaction rate increased by one-third. Therefore, with regard to O_2, increasing the concentration by one-third increases the reaction rate by one-third. (Please see the important note from the author at the beginning of the book.)

281. (D) All we need to do here is substitute the data from one of the trials into the rate law. Use the trial that has the easiest numbers to crunch. Using the data from Trial 1:

$$\text{Rate} = k[NO]^2[O_2]$$

$$k = \text{rate}/[NO]^2[O_2]$$

$$k = 0.143/[0.024]^2[0.035]$$

$$k = 7{,}093 \text{ or } 7.0 \times 10^3$$

Remember, the numerical value of the rate constant, k, always increases with increasing temperature. So the value of k calculated for a particular reaction occurring at one temperature will not be the same in the same reaction performed at a different temperature. (Please see the important note from the author at the beginning of the book.)

282. (C) We can memorize the units of the rate constant or calculate them. Notice that the absolute value of the exponent for L and mol are always one less than the reaction order.

Reaction Order	Units of k
1	sec^{-1}
2	$L\ mol^{-1}\ sec^{-1}$
3	$L^2\ mol^{-2}\ sec^{-1}$
4	$L^3\ mol^{-3}\ sec^{-1}$

To calculate k, just substitute the units into the rate law.

$$\text{Rate} = k[NO]^2[O_2]$$

$$M\ s^{-1} = k\ (mol\ L^{-1})^2\ (mol\ L^{-1})$$

Remember that the units of M are mol L^{-1}

$$mol\ L^{-1}\ s^{-1} = k\ mol^3\ L^3$$

$$k = L^2\ mol^{-2}\ sec^{-1}$$

(Please see the important note from the author at the beginning of the book.)

283. (C) The reaction order for NO is 2, so increasing its concentration would increase the reaction rate by 25 × ($[5x]^2 = 25$-fold increase in rate). (Please see the important note from the author at the beginning of the book.)

284. (D) All we need to do is plug the numbers into the rate law, using the value of k determined in **Question 281**.

$$\text{Rate} = k[NO]^2[O_2]$$

$$\text{Rate} = 7.3 \times 10^3\ [2 \times 10^{-2}]^2\ [4 \times 10^{-2}]$$

$$\text{Rate} = \mathbf{1.1 \times 10^{-1}}$$

(Please see the important note from the author at the beginning of the book.)

285. (E) The effect that the concentration of a particular reactant has on a reaction cannot be determined until the actual experiment of changing the concentration of that reaction and measuring the rate change that results is performed. In writing the equilibrium expression, the exponents that accompany reactions are taken from the balanced equation, but the equilibrium concentrations must be measured to actually determine the K_{eq} for the reaction. (Please see the important note from the author at the beginning of the book.)

286. (D) We should certainly remember the statements in choices (A), (B), (C), and (E) as important facts about catalysts. Choice (D) is the opposite of what is true, and this is also important to remember: A catalyst *does not differentiate* between the forward and reverse reactions. The activated complex is the same whether a reaction occurs in the forward or reverse reaction, so decreasing the energy of the activated complex will decrease the activation energy in both directions by an equal amount. Because a catalyst increases the rates of both forward and reverse reactions equally, the presence of a catalyst *does not change* the K_{eq} for a reaction. It doesn't change the ΔH_{rxn}, either (yet another important thing to remember about catalysts). (See **Answer 11** for an *except* question strategy.)

287. (E) Statements I and II are standard ways of measuring reaction rates. The rate law (III) can be used to predict the rate of a chemical reaction. The reaction quotient (Q) can be used in conjunction with the equilibrium constant (K_{eq}) to calculate how far a reaction is from equilibrium.

288. (A) Reaction rates are typically measured in one of two ways, the appearance of product over time, and/or the disappearance of one or more substrates over time. When using reaction data, it is vital to consider the stoichiometric coefficients when to compare relative rates of appearance or disappearance. For example, for each mole NO consumed by this reaction, one mole of H_2 will be consumed as well and one mole of H_2O will form, as well. However, only half of a mole of N_2 will form. Related to time, N_2 will be formed at half the rate as H_2O formation, and half the rate at which NO and H_2 are consumed.

289. (B) The slowest (rate determining) step in a multistep reaction governs the rate law of the reaction. CO is not involved in the rate determining step, so it is not included in the rate law. Because two NO_2 molecules are necessary for collision (it is bimolecular), the reactant order for NO_2 is two.

290. (D) It is more difficult to determine the rate law from a reaction mechanism in which an intermediate (NOBr, in this case) is involved in the slow (rate determining) step. Notice that the first, fast step is at equilibrium, so that the forward *and* reverse reactions are occurring at the same rate. We can write and equate a rate law for these reactions, then solve for [NOBr], which allows us to use the intermediate that we normally don't include in the rate law, but need when it is part of the rate determining step.

$$k_{forward} [NO][Br_2] = k_{reverse} [NOBr]$$

$$\mathbf{[NOBr] = k_{forward}/k_{reverse} [NO][Br_2]}$$

The rate law for the slow step, if NOBr was not an intermediate, would look like this:

$$Rate = k [NOBr][NO]$$

but now we substitute for [NOBr]

$$Rate = k_2 \mathbf{(k_f/k_r)[NO][Br_2]}[NO]$$

and simplify to get rate = $k_{rxn} [NO]^2[Br_2]$.

The value of $k_{rxn} = k_2 k_f / k_r$.

291. (D) The numerical value of the rate constant, k, is irrelevant to answering this question. It is the *unit* of the rate constant that gives us the most information (at first). From the unit, we can deduce that this is a third-order reaction. Remember, the absolute value of the exponent for L and mol is 1 less than the reaction order. (For reaction order *n*, the units of the rate law will be $L^{(n-1)} mol^{-(n-1)} sec^{-1}$.) (See **Answer 282** for a table of the rate constant units and how to determine them.)

The rate constant represents an "adaptor" between the concentration of reactants and the reaction rate (at a particular temperature), it doesn't tell us anything about the enthalpy changes that occur in the reaction, nor can it be used for comparing reaction rates with other reactions for which nothing but the rate constant is known. The units of the rate constant *do* provide a quantitative relationship between concentration changes and reaction rates. In this third-order reaction, doubling the concentrations of all the reactions will result in an eightfold increase in reaction rate ($[2x]^3 = 8 \times$ rate), *not* an 8.1×10^{10}-fold rate increase.

292. (B) Reactant orders can be positive, negative, or fractional. A negative reactant order indicates that the reaction rate increases as the reactant with the decreased concentration of that reactant.

293. (D) The activated complex formed during a chemical reaction without a catalyst *must* be different from the one formed by the same reaction *with* a catalyst. First, the potential energies are different. Second, the presence of a catalyst provides an alternative pathway for reaction, which means a different set of intermediates are formed. (See **Answer 11** for an *except* question strategy.)

294. (E) The reaction is endothermic because the products have more potential energy than the reactants. The E_a is calculated by subtracting the energy of the reactants from the energy of the activated complex ∴ 510 – 75 = 435 kJ mol^{-1}. The potential energy of the products is approximately 250 kJ mol^{-1} and the energy of the reactants is about 75 kJ mol^{-1} ∴ $\Delta H = 250 - 75 = 175$ kJ mol^{-1}.

295. (D) Reaction rate is measured by product formation (or substrate consumption) over time and is a function of the number of effective collisions that occur in a given period of time.

296. (B) It takes between 10 and 15 minutes to decrease the amount of reactant by 50 percent. It takes another 10 to 15 minutes to decrease the amount of reactant by another 50 percent, therefore the reaction order is 1, and the half-life is between 10 and 15 minutes ∴ 12 minutes.

297. (E) The total reaction order is 3, so doubling *both* reactants will increase the reaction rate eightfold ($([2x]^3)$ = 8-fold increase in rate).

298. (B) In Trial 2, the concentration of X is cut by half, so we expect the rate to decrease by half since X is a first-order reactant (∴ $^R/_2$). The concentration of Y is doubled, however, and Y is a second-order reactant, so the change in reaction rate due to doubling [Y] = 4R. $(^R/_2) \times (4R) = 2R$.

299. (D) In Trial 2, the concentration of X is cut by half so we expect the rate to decrease by one-fourth since X is a second-order reactant in this situation (∴ $^R/_4$). The concentration of Y is doubled, however, and since Y is a first-order reactant in this case, the change in reaction rate due to doubling is [Y] = 2R. $(^R/_4) \times (2R) = {}^R/_2$.

300. (D) The units of the rate constant for a fourth-order reaction is L^3 mol^{-3} sec^{-1}. Remember that the absolute value of the exponents of L and mol is 1 less than the reaction order. (See the table under **Answer 282** for a summary of the units of k.)

Chapter 8: Equilibrium

301. (D) *Le Chatelier's principle*: If a chemical system at equilibrium experiences a change in concentration, volume, temperature, or partial pressure, the equilibrium shifts to counteract the change and a new equilibrium will be established. To increase the amount of MgO produced, a disturbance to the equilibrium must be applied that will be counteracted by the production of *more* MgO. Removing $Mg_{(s)}$ (or adding it) will not cause any change in the equilibrium because it is a solid. Solids don't have a concentration and they are not represented in the equilibrium expression. *Increasing* the pressure on a gaseous system at equilibrium will cause the equilibrium to shift to favor the side of the reaction with the fewest moles of gas. *Decreasing* the pressure shifts the equilibrium to favor the side of the reaction with the most moles of gas. Changing the pressure of a gaseous system *does not change* the value of K_p, however. The ratio of partial pressures *after* the pressure change will equal the same value of K_p. Adding more $O_{2(g)}$ will increase MgO formation because the system will consume the O_2 in an effort to counteract the increase in O_2 pressure. (See table in **Answer 307**.)

302. (D) The reaction is endothermic because heat (energy) is on the reactant side of the equation. To shift the equilibrium to favor the reactants, the pressure could decrease (the reactants have greater number of moles of gas) or the temperature can be lowered (removing heat would pull the equilibrium to the left).

303. (E) This reaction is exothermic, so increasing the temperature favors the reactants by pushing the equilibrium to the left. The number of moles of gases on each side of the equation is the same, so pressure changes will not affect the equilibrium partial pressures.

304. (B) Increased pressure due to decreased volume shifts the equilibrium of a gaseous system to favor the side of the reaction with the fewest moles of gases (without changing the K_p). Equation (B) has an equal number of moles of gases in the products and reactants, so pressure changes will not change equilibrium partial pressures.

305. (D) Decreased water temperature and increased gas pressure above the water increase the solubility of a gas in water. Shaking the container will not result in a sustained increase in solubility (since the system is at equilibrium).

306. (B) If N_2 is injected into the tank, it will react with H_2 by consuming it, and therefore decreasing its partial pressure in the tank.

307. (C) The reaction is endothermic. The value for Kp *only changes with temperature changes.*

	Temperature ⇑	Temperature ⇓
Endothermic rxn (+ΔH)	K_{eq} ⇑	K_{eq} ⇓
Exothermic rxn (−ΔH)	K_{eq} ⇓	K_{eq} ⇑

308. (C) The ratio is the equilibrium expression for the reverse, exothermic reaction. *Only temperature changes can change the value of K_{eq}.* Pressure changes in a gaseous system will *shift* the equilibrium until the new equilibrium is reached. Although the actual partial pressures of each gas (at equilibrium) may be different, their ratio as calculated by the equilibrium expression will result in the same value of K_p.

309. (C) All we need to do is substitute the concentrations into the equilibrium expression.

$$K_{eq} = {}^{[\text{products}]}/_{[\text{reactants}]} = ([Y]^3\,[Z]) \div ([X]^2) = (125 \times 2) \div (16) = \sim 16$$

310. (C) The reactant side of the equation has fewer moles of gas than the products, so increasing the pressure will shift the equilibrium to favor the reactants, not the products. The value for ΔH is positive, so the forward reaction is endothermic, which means that increasing the temperature will shift the equilibrium to favor the forward reaction, and decreasing the temperature favors the reverse reaction. Adding more X will increase the formation of Z as the system consumes it in an effort to counteract the increase in X pressure.

311. (B) This question is not just asking what will shift the equilibrium, it's asking what will *change* the equilibrium constant for the reaction. The K_{eq} of a reaction doesn't change if we add more reactant or product, it only changes with temperature. The ratio in the question is written for the forward, endothermic reaction (the reverse reaction would be the inverse). (See table in **Answer 308**.)

312. (E) The equation tells us that for every *one* F_2 that reacts, *two* F are produced. If we put the number of particles in the box into the equilibrium expression we get $[F]^2/[F_2] = [4^2/9] = 1.8$. The equilibrium constant will never be a negative number because there are no negative concentrations.

313. (E) The equilibrium expression for the reverse reaction inverts the products and reactants, so the K_{eq} for the reverse reaction is simply the inverse, $1/K_{eq}$. The value of $1/2.5 \times 10^{10}$ is easy to calculate because all we need to do is divide 1 by 2.5 (= 0.4) and reverse the sign on the exponent (10^{-10}). However, 0.4×10^{-10} is more correctly written as 4.0×10^{-11}.

314. (D) If 30 percent of XY_3 dissociates in 1 kg of a 0.06 *m* solution, then out of the 0.06 mole of XY_3 present, 0.018 mole dissociates and 0.042 mole remains in the form of XY_3. Immediately, choice (A) can be eliminated. Each XY_3 that dissociated produced 4 particles, one X^{3+} and three Y^-, so when the 0.018 mole of XY_3 dissociated, it created $0.018 \times 4 = 0.072$ mole of particles. The total particle count, then, is 0.042 mole undissociated particles plus 0.072 mole ions = 0.114 mole particles.

315. (C) K_{sp} is the solubility product constant, the equilibrium constant for a solid in equilibrium with its ions. It is measured in a saturated solution at equilibrium. The molar solubility is the number of moles that can be dissolved to produce 1 L of a saturated solution of that substance. What simplifies K_{sp} calculations is that there is always a solid reactant in these problems but *solids are not included in equilibrium expressions, so they will never have a denominator.*

$$BaSO_{4(s)} \Leftrightarrow Ba^{2+}{}_{(aq)} + SO_4{}^{2-}{}_{(aq)}$$

$$K_{sp} = [Ba^{2+}]\,[SO_4{}^{2-}]$$

Let $x = [Ba^{2+}] = [SO_4{}^{2-}]$ (They are produced in a 1:1 ratio.)

$$K_{sp} = x^2 = 1.1 \times 10^{-10}$$

$$x = \sim 1 \times 10^{-5}$$

(See **Answer 316** for a more complicated variation of this problem.)

316. **(E)** $CaF_{2(s)} \Leftrightarrow Ca^{2+} + 2\ F^-$

Let $x = [Ca^{2+}]$

F^- and Ca^{2+} are produced in a 2:1 ratio, so $[F^-]$ is $2x$. Because K_{sp} is an equilibrium constant, the $[F^-]$ must be raised to the exponent that is its stoichiometric coefficient, 2.

$$K_{sp} = [Ca^{2+}]\ [2\ F^-]^2$$

$$K_{sp} = 4x^3$$

$$4.0 \times 10^{-11} = 4x^3 \therefore x^3 = 1.0 \times 10^{-11} \therefore x = (1.0 \times 10^{-11})^{1/3}\ M$$

Remember that the $[Ca^{2+}] = x$, but the $[F^-] = 2x$.

(See **Answer 315** for a simpler variation of this problem.)

317. **(C)** This is the reverse process of **Questions 315 and 316**, but we'll use the same principles. There is also an extra step involved because instead of being given a value for the molar solubility of one of the dissociation products (given which we can deduce the concentrations of the other), we are given the pH. The pH = 10, so we know the solution is basic.

$$pH + pOH = 14, \text{ if } pOH = 4 \text{ then } [OH^-] \text{ is } 1 \times 10^{-4}\ M$$

$$M(OH)_2 \Leftrightarrow M^{2+} + 2\ OH^-$$

Let $x = [M^{2+}]$

$$[OH^-] = 2x = 1 \times 10^{-4} \therefore x = 0.5 \times 10^{-4}, \text{ or } 5 \times 10^{-5}\ M$$

$$K_{sp} = [M^{2+}]\ [2\ OH^-]^2 \therefore K_{sp} = [x]\ [2x]^2 = 4x^3$$

Since $x = 5 \times 10^{-5}$, $4x^3 = 5 \times 10^{-13}$.

318. **(A)** The solubility of $Fe(OH)_2$ decreases with increased pH because the higher concentrations of OH^- ions pushes the reaction toward the formation of solid $Fe(OH)_2$. Decreasing the pH *increases* its solubility because the extra H^+ ions in solution will combine with the hydroxides to produce H_2O, pushing the reaction to the left. The value of K_{sp} does not depend on the pH directly, but since one of the aqueous products of the reaction is OH^-, the K_{sp} already takes the pH into account.

319. **(A)** The K_{sp} for $BaCrO_4$ is much smaller than that of $CaCrO_4$, so we know that it is much less soluble. The difference of six orders of magnitude (10^{-10} versus 10^{-4}) tells us that it is possible to precipitate almost *all* of the Ba^{2+} ions without precipitating any $CaCrO_4$. The trick to solving this problem is to understand how to produce a saturated $CaCrO_4$ solution. This will leave Ca^{2+} ions in solution, but will overwhelm the Ba^{2+} ions.

$$K_{sp} = [Ca^{2+}][CrO_4^{2-}]$$

$$7 \times 10^{-4} = [0.2][CrO_4^{2-}] \therefore [CrO_4^{2-}] = \sim 3.5 \times 10^{-3}\ M.$$

320. (E) The only determinant of the vapor pressure of a particular substance at equilibrium with its liquid is the temperature. The amount of surface area will affect the rates of evaporation (and condensation) and the volume of the container will affect how many moles of vapor are present, but not the *vapor pressure at equilibrium.*

321. (C) The vapor pressure of a liquid is determined at equilibrium at a particular temperature, so we can consider this as we would an equilibrium question. Just as the K_{eq} of an endothermic reaction *increases* with increasing temperature (and the K_{eq} of exothermic reactions *decreases* with increasing temperature), higher temperatures favor vaporization over condensation until a new equilibrium is reached, and it will include a larger fraction of vapor than the equilibrium condition at a lower temperature. Choice (A) is not true and the system in choice (B) is not at equilibrium. For a system at equilibrium (and at constant temperature), the rates of forward and reverse processes are the same, so (D) is not true. Finally, the ΔH of a forward and reverse process (or reaction) are equal in magnitude, but opposite in sign.

322. (C) At atmospheric pressures, N_2, O_2, and Ar condense at −195°C, −183°C, and −186°C respectively. CO_2 does not condense into a liquid at atmospheric pressures, but becomes a solid at about −80°C (see the phase diagrams for H_2O and CO_2 above **Questions 96 and 99**, respectively).

323. (A) An adiabatic system is one in which heat does not enter or leave the system. Entropy is a measure of the degree of disorder or dispersion of energy in a system, and it typically increases with the addition of heat. The liquid state of a pure substance has greater entropy than its solid state because the particles in a liquid are more disordered than the particles in a solid.

During the melting process, the change in entropy is equal to the heat of fusion divided by the melting point in K. The term melting implies a displacement from equilibrium by continued heat transfer over time to drive the melting process. However, in this question, the system is adiabatic, that is, no heat is entering or leaving the system, it is at a constant temperature.

The phase equilibrium of a pure substance is independent of the ratio of the solid phase to the liquid phase. Equilibrium exists even if only the tiniest bit of solid is present in the solid–liquid mixture. For that portion of the solid that melted, there is an entropy increase. The remaining mass of liquid and solid is at equilibrium and has no entropy change.

324. (B) The strength of oxyacids increases with (1) increasing electronegativity of the central atom (the atom the O—H attaches to), (2) other atoms of high electronegativity in the compound, and (3) an increasing number of oxygen atoms. The pH of a solution depends on the concentration on H^+ in solution and therefore the concentration of the acid in solution, but *the strength of an acid is intrinsic to the acid and does not depend on the concentration.*

325. (C) The K_a for hypoiodous acid is 2×10^{-11}. The stability of the conjugate base also tells us something about the strength of the acid OI^-, which picks up protons in solution rather easily, meaning that more HOI is present at equilibrium. If less HOI dissociates, the acid is weaker. (See **Answer 324** for a list of description of factors that affect the strength of oxyacids.)

326. (C) The names of oxyacids are derived from the polyatomic anions that the protons are attached to.

In the case of ClO^- series:

ClO^- = **hypo**chlor**ite**

ClO_2^- = chlor**ite**

ClO_3^- = chlor**ate**

ClO_4^- = **per**chlor**ate**

The ClO^- ion is named hypochlorite because it has the least number of oxygen atoms in the series. The compounds are named relative to each other, *not* by their absolute number of oxygen atoms. Once we've identified the polyatomic oxyanion, we name the acid accordingly:

Oxyanions ending in *–ite* become *–ous acid.*

Oxyanions ending in *–ate* become *–ic acid.*

327. (D) The equation shows the reaction of the conjugate base of HOCl. The formation of an OH^- ion means we can use K_b, the base ionization constant, to figure out the unknowns in the reaction.

$(K_a)(K_b) = K_w = 1.00 \times 10^{-14} \therefore K_b = 1.00 \times 10^{-14} \div K_a$

328. (D) The AP Chemistry exam is unlikely to give us a question like this in the multiple choice section, however there is a mandatory equilibrium problem in the free response section, so we need to know how to set up an ICE-box. See choice (D). (1) The initial concentration of reactants is taken from the question. Water is a pure liquid and is therefore *not considered* in the equilibrium expression. (2) The initial concentration of both products is zero. (3) The change in concentrations will always be (–) for the reactants and (+) for the products, *but their stoichiometric coefficients from the balanced equation MUST be accounted for.*

For example, if 2 A → 3 B, then the change in [A] = $-2x$ and the change in [B] = $+3x$.

329. (C) The rate of the forward reaction continues to decrease until about 1 second. The instantaneous reaction rate is determined by calculating the slope of the tangent to the point of interest on the concentration versus time curve. At point X, the rate of the forward reaction is still decreasing and the rate of the reverse reaction is still increasing. At point X on this graph, the slope of curve A is the inverse of the slope of curve B.

330. (D) Le Chatlier's principle says that when a chemical system at equilibrium experiences a change in concentration, the equilibrium shifts to counteract the change imposed on it. Once the system regains its equilibrium state, the ratio of products to reactants stays the same (only temperature changes affect the K_{eq} of the reaction).

Chapter 9: Acid–Base Chemistry

331. (B) Definition of a Bronsted-Lowry acid.

332. (D) Definition of a Lewis base.

333. (E) Definition of an Arrhenius base.

334. (D) Coordinate covalent bonds are covalent bonds in which one atom donates *both* electrons to the bond, instead of each atom donating one electron. There is no difference in the character of the bond, only how it forms. Lewis bases always form coordinate covalent bonds with acids since Lewis bases are *electron pair donors.*

335. (B) Choice (B) is a mixture of a strong base and a weak base.

336. (C) Choice (C) is a mixture of a strong acid and a neutral salt. The Cl^- ion is a very poor conjugate base and will not pick up any protons from the solution, so the pH is 0. (A good fact to memorize: The pH of a 1 M concentration of a strong acid = 0.)

337. (A) A weak base and a weak acid will neutralize each other.

338. (E) An alkaline buffer consists of a weak base and the salt of its conjugate acid. A buffer resists changes in pH by acting like a "proton sponge," giving them to the solution when the $[H^+]$ decreases, and absorbing them from the solution when the $[H^+]$ increases.

339. (D) An acidic buffer consists of a weak acid and the salt of its conjugate base. A buffer resists changes in pH by acting like a "proton sponge," giving them to the solution when the $[H^+]$ decreases, and absorbing them from the solution when the $[H^+]$ increases.

340. (B) We can use the expression for K_a.

$$K_a = 4.5 \times 10^{-4} = [NO_2^-][H^+] \div [HNO_2]$$

Let $[H^+] = x = [NO_2^-]$

Let $[HNO_2] = 0.02 - x$ but because the value of K_a is so low and we will not be calculating the value of the $[H^+]$ to more than two or three significant figures, the value of $(0.02 - x)$ is indistinguishable from 0.02, so we can drop the x in the denominator and greatly simplify our math.

$$\therefore 4.5 \times 10^{-4} = x^2/0.02 = 0.003 \text{ or } 3 \times 10^{-3}$$

341. (E) In the reaction, F^- is the proton acceptor, the base, and H_2O is the proton donor, or acid. When dealing with *Bronsted–Lowry acids and bases,* the acid and base are always the reactants, and their conjugates are always the products. The base, once it accepts an H^+, becomes the conjugate acid and the acid, once it donates an H^+, becomes the conjugate base.

342. (A) When dealing with *Bronsted–Lowry acids and bases,* the acid and base are always the reactants and their conjugates are always the products. Since H_2O is a reactant, it can only be an acid or base, not a conjugate. Be careful with questions that ask about *the reverse reaction.* For example, water *is* the conjugate acid of the OH^- ion in the reverse of this reaction. But in the forward reaction, H_2O is *giving the H^+ to NH_3, so it is an acid.*

343. (E) An increased number of hydrogens does not make an oxyacid a stronger acid. If an acid has more than one dissociable proton, it is a *polyprotic acid.* The K_a of the first H^+ is the highest, and the K_a decreases for each successive H^+. (See **Answer 324** for a list of factors that affect the strength of oxyacids.) (See **Answer 11** for an *except* question strategy.)

344. (E) When mixing two solutions, only concentrations intermediate to the solutions being mixed can be achieved. In this case, we can't make a solution that is less concentrated than 0.25 M and we can't get more concentrated than 0.35 M, so all concentrations achievable by mixing these two solutions, no matter what the ratios, will be between these two values.

345. (C) The K_a for the acid HA is 5.0×10^{-9}. A K_a of such small magnitude means the acid is very weak. Compared to 0.5 M concentration, the amount that dissociates is almost negligible in calculations in which the answer will contain two to three significant figures. This makes these calculations much easier because you can ignore the value of x in the denominator, as follows. Start by setting up the chemical equation and the equilibrium expression for the value of K_a:

$$HA \Leftrightarrow H^+ + A^-$$

Let $x = [H^+]$ $\therefore$ $x = [A^-]$

$$5.0 \times 10^{-9} = [H^+]\,[A^-]/[HA]$$

$5.0 \times 10^{-9} = x^2/[0.5 - x]$ but the value for x is so low we can ignore it, so essentially,

$$5.0 \times 10^{-9} = x^2/2$$

$\therefore$ $x = [H^+] = 1 \times 10^{-4}$ M. The negative log of 1×10^{-4} is 4 $\therefore$ the pH is 4.

346. (B) An equilibrium constant of 2.5×10^3 is large, and so the products are greatly favored in this reaction. Y^- is the proton acceptor in the forward reaction and X^- plays that role in the reverse reaction. But since the forward reaction is favored so much more than the reverse, we can infer that Y^- is a stronger base than X^-. We can also infer that HX is a stronger acid than HY, but that is not an answer choice. As for conjugate acids and bases, X^- is the conjugate base of HX, and HY is the conjugate acid of Y^-. A conjugate acid will differ from its base by one more H^+. A conjugate base will differ from its acid by one less H^+.

347. (D) The equivalence point where the number of moles of base (or acid) added equals the number of moles of acid initially present (or base). It is midway along the steep part of the curve, *the inflection point.* The pH at the equivalence point will be 7 for a strong acid/strong base titration, but will be greater than 7 when titrating a weak acid with a strong base. It will be below 7 when titrating a weak base with a strong acid. This is because the equivalence point *is not* when the actual number of OH^- ions is equal to the actual number of H^+ ions; it is, in this case, when the number of OH^- ions *added* equals the number of H^+ ions *initially present*, that is, neutralization.

348. (B) The region of the curve between 0 and about 25 mL NaOH added is the buffer region. At the lower end of the buffer region, the amount of weak acid exceeds that of the conjugate base and toward the end, the conjugate base is present in much greater concentrations. Midway, the weak acid/conjugate base concentrations are about equal.

349. (A) A buffer resists changes in pH by acting like a "proton sponge," giving them to the solution when the $[H^+]$ decreases, and absorbing them from the solution when the $[H^+]$ increases. A buffer consists of an acid or base and the salt of its conjugate base or acid. The region of the curve between 0 and about 25 mL NaOH added is the buffer region. The weak acid and its conjugate base are both present and very little neutralization has occurred.

350. (D) The question is asking us to identify the compound that would produce a basic solution. LiF is a soluble salt (any ionic with an alkali metal is soluble) that produces F^- upon dissociation. F^- is a good conjugate base and readily picks up protons in solution, increasing the pH.

351. (E) Conjugate acid/base pairs always differ by *one proton* (H^+). If we are looking for a conjugate acid, it means that NH_3 is acting as a base, accepting a proton. The conjugate acid would be NH_4^+, one more H^+.

352. (B) Two handy rules regarding pH and pOH are:

$$\mathbf{pH + pOH = 14}$$

$$\mathbf{[H^+][OH^-] = 1 \times 10^{-14}}$$

This problem can be solved in two ways. (1) If the pH = 6, the pOH = 8 $\therefore$ the $[OH^-] = 1 \times 10^{-8}$. (2) If the pH = 6, the $[H^+] = 1 \times 10^{-6}$ $\therefore$ the $[OH^-] = 1 \times 10^{-8}$. This makes sense because we expect the $[OH^-] < 1 \times 10^{-7}$ in an acidic solution (pH 6).

353. (B) To answer this question correctly we must remember that the pH scale is a $\log_{10}$ scale, so a pH of 6 is 10 times more acidic than a pH of 7 and 100 times more acidic than a pH of 8. If we want to acidify the pH by 1 point on the pH scale, we must either reduce the $[OH^-]$ by 10 and increase the $[H^+]$ by 10 (an increase in one will always result in a decrease of the other, see the two rules in **Answer 352**). So we must dilute the basic solution tenfold to decrease the pH from 13 to 12.

354. (E) The values for K_1, K_2, and K_3 decrease successively, so we expect the anion of the third deprotonation to be present in the lowest concentration. We can also consider that only a small fraction of $H_3C_6H_5O_7$ will lose a proton, leaving $H_2C_6H_5O_7^-$ behind. Let's say 1 percent to make it simple, although it's actually much lower than that. Out of the small number of $H_2C_6H_5O_7^-$ ions, only a small fraction of those will lose the second proton, leaving $HC_6H_5O_7^{2-}$ behind. Again, let's say 1 percent for simplicity. That means that 1 percent of 1 percent (0.01 percent) of the original $H_3C_6H_5O_7$ loses two protons. Finally, if 1 percent of $HC_6H_5O_7^{2-}$ loses the third proton, only 1 percent of 0.01 percent of the original $H_3C_6H_5O_7$ would have lost all three protons (0.0001 percent).

355. (C) A buffer resists changes in pH by acting like a "proton sponge," giving them to the solution when the $[H^+]$ decreases and absorbing them from the solution when the $[H^+]$ increases. A buffer consists of an acid or base, and the salt of its conjugate base or acid. Choice I represents a weak base and its conjugate acid, and choice III represents a weak acid and its conjugate base. Choice II, however, represents a strong acid and the salt of its conjugate base. The conjugate bases of strong acids are very weak and are not able to pick up protons in solution, so they are not good buffers.

356. (E) 0.2L × 0.2 mol/L = 0.04 mole $Sr(OH)_2$ but × *two* OH^- per $Sr(OH)_2$, so 0.08 mole OH^-

$0.8 \times .8 \times 2 = 1.28$

1.36 M OH = pH of 14.

357. (D) The final volume is 200 mL. We can use $M_1V_1 = M_2V_2$ to solve this problem: $(100)(0.002) = (M_2)(200) = 0.001$ M, or just realize that the concentration has been halved by adding an equal volume of water to the solution. HCl is a strong acid and so the concentration of HCl is equal to the $[H^+]$. $pH = -\log [H^+] \therefore -\log [1 \times 10^{-3}] = 3$.

358. (D) The final volume will be 20 mL. We can use $M_1V_1 = M_2V_2$. $(5)(0.02) = (M_2)(20) = 0.005$ M *but* that is the molarity of the $BaOH_2$ solution, *not* the $[OH^-]$ concentration. There are *two* OH^- per $BaOH_2$ so the concentration of OH^- is two times that of the $[BaOH_2] \therefore 0.01$ M. The $pOH = -\log [OH^-] = -\log [1 \times 10^{-2}] = 2 \therefore pH = 12$.

359. (C) Phenolphthalein changes from clear in acid to pink in base. We add the phenolphthalein to the acid and titrate with the base. When the acid we are titrating turns pink, we know the endpoint has been reached.

360. (C) The numbers on the graduations differ on the two burettes, so we need to be careful to read both burettes and not just take the difference between the heights of them. Since it's the same solution, we can read from the meniscus *or* we can read from the edge. It doesn't matter in a question like this because we'll get the same answer either way. We will read from the edge because it's closest to a whole number, 10, on the first burette, and 41.5 on the second burette.

The change in volume $= V_{final} - V_{initial} = 41.5 - 10 = 31.5$ mL.

361. (C) We use $M_aV_a = M_bV_b$, but we need to remember that $Sr(OH)_2$ has *two* moles of OH^- ions per mole $Sr(OH)_2$. $[(M_a)(35) = (0.5 \times 2)(25)] \therefore M_a = \sim 0.7$ M.

Chapter 10: Electrochemistry

362. (B) The answer to this question is very clear if we know a simple fact: Zn takes on only *one* oxidation state, +2. (A trick for remembering the oxidation states of Ag, Zn, and Al: they are connected diagonally on the periodic table, and as we progress from Ag to Zn to Al, the oxidation states are +1, +2, and +3.)

363. (E) The sum of all the oxidation states of all the atoms must sum to the overall charge of the compound. In this case, the oxidation state of perchloric acid is zero:

$$H = +1 \text{ and } O = -2\ (\times 4) = -8\ \therefore\ Cl = +7$$

364. (E) The sum of the oxidation numbers on the atoms in the chromite ion, $Cr_2O_4^{2-}$, must equal the charge on the compound, −2. $O = -2\ (\times 4) = -8\ \therefore\ 2\ Cr = +6\ \therefore\ 1\ Cr = +3$. The answer choices provide some complicated compounds for assigning oxidation numbers. Try the easiest ones first. Cr_2O_3 looks promising because it only has three oxygen atoms and no charge. The −2 charge on $Cr_2O_4^{2-}$ negated the effect of the extra oxygen, so the Cr atom in Cr_2O_3 should be the same (and it is). Chromium takes on oxidation states ranging from −2 to +6. Negative oxidation states are not common for metals. $Cr_2(O_2CCH_3)_4$ is chromium (II) acetate, and features a *quadruple bond* (which occurs extremely rarely). $Cr(CO)_6$ is chromium hexacarbonyl, a compound in which Cr takes on a zero oxidation state in this compound. The Cr in dichromate, $Cr_2O_7^{2-}$, has an oxidation state of +6, and potassium peroxochromate, $K_3[Cr(O_2)_4]$, features a +5 oxidation state for Cr.

365. (E) The electrical conductivity of a solution is determined mainly by the concentration of its ions and its temperature. The concentration of ions in solutions equimolar with respect to compounds varies according to solubility and the van't Hoff factor (i). *Strong acids and bases, and soluble salts are good electrolytes because they dissociate completely.* H_3PO_4 is a weak acid and therefore does *not* dissociate into 3 H^+ and a PO_4^{3-} as we might expect; however, if dissolved in a basic solution, it can. The K_a values for the protons are 7.5×10^{-3}, 6.2×10^{-8}, and 4.8×10^{-13}. For a given acid in solution, if the pH < pKa "the proton is on" and if the pH > pKa, "the proton is off."

366. (A) Some oxidation and reduction reactions are spontaneous (exergonic). Our table of Standard Reduction Potentials is an incomplete list of them. Reductions that have a positive E° value are exergonic, which means the reverse reaction, or oxidation, is endergonic. The bottom of the table would have endergonic reductions and exergonic oxidations. Notice reducing F_2 is very exergonic and reducing Li^+ is very endergonic. The reduction of (1 M) Cl_2 produces 1.36 V, while the oxidation of (1 M) Br^- requires 1.07 V. The difference between them, 0.29 V, is still favorable (+ΔE = spontaneous) so this redox reaction is spontaneous and can generate electricity. We can think of this in terms of how much Cl_2 "wants" to be reduced versus how much Br^- "doesn't want" to be oxidized. The difference in the ΔE° values quantifies how greatly their "desires" differ and which species will get what it wants. Comparing the E° values for the two elements tells us that the ability of Cl_2 to take electrons from Br^- is greater than the ability of Br^- to prevent Cl_2 from taking them.

367. (C) Statement I is not true because Cu starts out in the solid, reduced form and ends up as a 2+ ion, so it is certainly *not* the oxidizing agent. (*Agents* get the opposite of what they do. If the agent *oxidizes*, it gets *reduced.*) Statement II is also not true. The oxidation state of hydrogen does not change in the reaction. It is +1 as an ion and +1 in water.

368. (C) We almost certainly have heard the mnemonic device that applies to all electrochemical cells (galvanic *and* electrolytic), **An Ox, Red Cat,** which translates to ***An**ode* = ***Ox**idation, **Red**uction at the **Cat**hode.* The two reactions are written as reductions, and so one of them will need to be written in reverse. Since they are opposite signs, we should flip the nonspontaneous reaction, the one with the $-\Delta E°$. Silver is reduced at the cathode producing 0.80 V, and Zn is oxidized at the anode, producing 0.76 V. Total cell potential = 1.56 V. The Zn^{+} ion does not participate in this reaction.

369. (D) See explanation for **Answer 368.**

370. (D) See **Answers 363 and 364** for examples of how to assign oxidation numbers.

371. (C) Aluminum gets reduced (its oxidation state changes from +3 to 0) and oxygen gets oxidized (its oxidation state changes from −2 to 0). **An Ox, Red Cat,** which is ***An**ode* = ***Ox**idation, **Red**uction at the **Cat**hode.*

372. (D) An ampere quantifies the rate at which charge (measured in Coulombs, C) flows in C/sec. The charge on *one* proton or electron is +/− 1.6×10^{-19} C (the +1 and −1 we use in other chemistry chapters refer to an elementary charge, the numerical value of which is +/− 1.6×10^{-19} C). A *faraday* is the magnitude of the electrical charge on one mole of electrons (or protons). A current of 15 amperes means 15 C, or 15 mole electrons, are available per second (pass into the cell), for 20 minutes (1,200 seconds). The numbers 20 and 60 in the numerator convert minutes to seconds. The number 27 converts grams of Al into moles.

The number 15 in the numerator converts the amperage and time into Coulombs: amps ($^{C}/_{s}$) × time (sec, from the product of 20 × 60) = Coulombs. 1 mole e^{-} = 96,500 C, so if 18,000 C are 0.187 mole e^{-} enter the cell. Each mole of Al requires 3 moles of e^{-} to get reduced, so 0.187 ÷ 3 = 0.062 mole of Al will get reduced. At 27 g mol^{-1}, that's 1.7 g Al. An interesting fact about aluminum is that it's the third most abundant element on earth, but its electrolytic production is very energy intensive. Approximately 4 percent of the electricity consumption in the United States is used for the production of aluminum.

373. (D) The reduction of X^{2+} to X is the reverse of Reaction 1, so the DE° of the reaction is −2.27 V, but the oxidation of Ag also occurs. Reaction 2 gives us the reduction of Ag^{+}, which produces 0.80 V, so its oxidation is −0.8 V. Subtracting the oxidation of Ag (−0.80 V) from Reaction 1 (−2.27 V) leaves us with the half-reaction for the reduction of X and a standard reduction potential of −1.47 V.

374. (D) An ampere quantifies the rate at which charge (measured in Coulombs, C) flows in one second, so the unit of current is $^{C}/_{sec}$. The charge on *one* proton or electron is +/− 1.6×10^{-19} C (the +1 and −1 we use in other chemistry chapters refers to an elementary charge, the numerical value of which is +/− 1.6×10^{-19} C). A *faraday* is the magnitude of the electrical charge on one mole of electrons (or protons).

1.00 ampere = 1 C/sec. The oxidation state on Fe in $FeCl_3$ is +3, so it will require 3 moles of e– to reduce 1 mole Fe^{3+}.

3 moles of e^- = 96,500 C × 3 = 289,500 C required. Delivered at a rate of 1.00 C/sec, that would require 289,500 seconds (more than 80 hours).

375. (B) A simple way to eliminate some answer choices is to count electrons. We need to remember we are dealing with the Zn^{2+} ion, which has *two* fewer electrons than Zn $\therefore 30 - 2 =$ 28 electrons. Choices (B), (D), and (E) have 28 electrons so we've already eliminated two choices. The electrons of highest energy (n) are lost first, so the Zn atom loses its 4s electrons leaving a full d subshell. That leaves only choice (B).

376. (D) The number of positive charges in the nucleus of Zn^{2+} is still 30, but now those 30 protons are only pulling on 28 electrons.

377. (A) Use the mnemonic **An Ox, Red Cat,** which comes to ***Anode*** = ***Oxidation, Reduction*** *at the* ***Cathode***. The question is, which species—Zn or H_2—gets oxidized. We may remember that in the hydrogen electrode, 2 H^+ are reduced to H_2, but if we don't, that's fine. We know the cell is spontaneous because it is a galvanic cell and therefore the E° of the cell must have a positive number. Since the reduction of Zn^{2+} is negative, the oxidation is positive, and our cell will work spontaneously. The hydrogen electrode is the cathode, so choices III and IV do not apply.

378. (D) The salt bridge has two functions in the galvanic cell. (1) It completes the circuit by providing a bridge of mobile ions between the two half cells and (2) it provides ions to the half cells extending the life of the battery.

At the anode, the oxidation of solid Zn produces Zn^{2+} ions in solution. The salt bridge neutralizes the excess positive charges by providing NO_3^- ions. At the cathode, H^+ ions are being reduced to H_2 gas, so the salt bridge provides K^+ ions to replace the lost cations, neutralizing the Cl^- ions.

379. (E) It's not possible to measure the E° of a half reaction (or half cell) so the reduction of 2 H^+ to H_2 was given the arbitrary value of zero and all the other standard reduction potentials were measured relative to it.

380. (C) We know *the cell is spontaneous because it is a galvanic cell,* not an electrolytic cell. Spontaneous cells have a *negative* ΔG and the sum of the oxidation and reduction potentials must be *positive.*

381. (E) The Nernst equation relates half-cell concentrations to the E of the cell by $E = E° - (^{0.0257\ V}/_{n}) \ln Q$, where Q is the reaction quotient. As the cell operates, the flow of electrons from the anode to the cathode results in the formation of product and the consumption of reactant. This increases product concentration and decreases reactant concentration. As the cell approaches equilibrium, E = 0 (and $Q = K_{eq}$). Zn^{2+} is the product of the oxidation that occurs at the anode.

382. (A) A sacrificial anode is also called a *galvanic anode*, and is a metal with a more negative electrochemical potential. This allows the sacrificial anode, the Zn, to be oxidized preferentially to the "cathode," in this case, the iron.

383. (B) Electrical energy = electrical potential × the base unit of charge = Voltage × Coulombs. A volt = 1 J C^{-1} ∴ electrical energy can be expressed in joules. The sum of the two half-cells only determines the maximum voltage of the cell, not how much energy the cell can generate.

384. (C) In an electrochemical wet cell, concentration *does* matter. The E° values on the table of Standard Reduction Potentials are for *standard conditions* that include 1.0 M concentrations. The Nernst equation is used to calculate the electromotive force (voltage) of a cell under *nonstandard conditions.* (See **Answer 11** for an *except* question strategy.)

385. (B) From the balanced redox equation we can see that $Li_{(s)}$ is oxidized, and then we remember: **An Ox, Red Cat:** *Anode* = ***Ox****idation,* ***Red****uction at the* ***Cat****hode.*

386. (E) Choices (A) through (D) make lithium-ion batteries superior to many other types of dry cells, but they are very reactive. They can explode if not handled with care. (See **Answer 11** for an *except* question strategy.)

387. (A) *All* electrochemical cells abide by **An Ox, Red Cat:**

Anode = ***Ox****idation,* ***Red****uction at the* ***Cat****hode.* However, galvanic (voltaic) cells are spontaneous and convert chemical energy into electrical energy whereas electrolytic cells require energy to cause nonspontaneous redox reactions (electrolysis).

Chapter 11: Nuclear Chemistry

388. (E) When balancing nuclear reactions, remember: (1) *Conservation of mass number* (the number of protons and neutrons are equal in the products and reactants) and (2) *Conservation of atomic number* (the number of nuclear charges in the products and reactants are the same).

Total mass number in reactants = 236

Total atomic number in reactants = 92

Mass number in products must = 236 ∴ 236 − 141− (3 × 1) = 92

Atomic number in products must = 92 ∴ 92 − 55 = 37, rubidium

389. (B) See **Answer 388**.

Total mass number in reactants = 31

Total atomic number in reactants = 15

Mass number in products must = 31 ∴ 31 − 1 = 30

Atomic number in products must = 15 ∴ 15 − 0 = 15, phosphorus

390. **(B)** See **Answer 388**.

Total mass number in products = 18

Total atomic number in products = 9

Mass number in reactants must = 18 ∴ 18 − 4 = 14

Atomic number in reactants must = 9 ∴ 9 − 2 = 7, nitrogen

391. **(D)** Choice (D) = gamma radiation (γ), a high-frequency form of electromagnetic radiation (EMR). Visible light is a very low frequency of EMR relative to gamma, but the basic properties of EMR are its wave- and particle-like properties, and its lack of mass.

392. **(A)** Choice (A) is alpha decay. An alpha particle is a helium nucleus, with a mass number of 4. It makes it the most massive particle and therefore least penetrative form of radioactive decay. (For a given value of kinetic energy, more massive particles have lower speed.)

393. **(D)** Choice (D) = gamma radiation (γ), the highest frequency of electromagnetic radiation (EMR) known. The frequency of EMR is proportional to its energy, so gamma radiation is the highest energy EMR known. The frequency of gamma radiation is on the order of magnitude of 10^{19} Hz, and the wavelengths are less than the diameter of an atom. Gamma radiation is ionizing because it has enough energy to dislodge electrons from atoms and molecules and produce free radicals in the body.

394. **(A)** Choice (A) is an alpha particle, a helium nucleus with a 2+ charge. Other charged decay particles have a +1 or −1 charge.

395. **(B)** Choice (B) is beta decay, a form of decay in which a neutron turns into a proton and electron. The electron is ejected from the nucleus as a beta-particle ($^{0}_{-1}e$ or $^{1}_{-1}\beta$), and the atomic number increases by 1, leaving the mass number intact (so the number of neutrons decreases by 1). Beta-decay *decreases* the neutron to proton ratio of an unstable nucleus.

396. **(C)** Nuclear binding energy is an indication of the stability of a nucleus. The binding energy is the energy required to break up a nucleus into protons and neutrons. The nucleus with the highest binding energy per nucleon is the most stable.

397. **(A)** Choice (A) is false because the curve represents *binding energy per nucleon*, not per nuclide. More importantly, the stability of a nucleus is the sum of *two* forces, attractive and repulsive. The *net attractive force* is what ultimately determines the stability of the nucleus, not the total attractive force (without considering the repulsive forces). (See **Answer 11** for an *except* question strategy.)

398. **(C)** The definition of mass defect. The binding energy is the energy required to break up a nucleus into protons and neutrons.

399. (D) The lost mass is accounted for by Einstein's famous mass–energy equivalence equation, $E = mc^2$. Since the value of $c^2 = 9 \times 10^{16}$, a tiny piece of mass lost from the nucleus is transformed into a lot of energy.

400. (E) Statement I correctly accounts for the chemical reactivity of the alkali metals. *Chemical reactivity* (or stability) is a function of *electron configuration,* whereas *radioactivity* is a function of *nuclear stability* (neutron-to-proton ratio). Statement II is *incorrect*; radioactive nuclei increase their stability by decaying, often transmuting into another species of atom. Statement III is correct about *all* the non-noble gas elements; they all form molecules that are more stable than the lone atoms of the element. The "motivation" of atoms is to fill their valence shell by giving, taking, or sharing electrons with other atoms.

401. (A) This is the strong nuclear force, and it only works when the protons very close to each other are on the order of about 10^{-15} m. The nucleus only carries positive charges (protons) although it can create and emit electrons (β-particles, ${}^{0}_{0}e^{-}$) through β-decay. The nucleus is dense, so the protons can't be spread out enough to escape the like-charge repulsion. The neutrons don't form a cage around the protons. There would need to be a force that holds the neutrons together in cage formation, and that force would need to exceed the repulsive force of the like-charge repulsion to be an effective cage. Besides the fact that it isn't true, it doesn't explain anything about the forces; it only implies a new force exerted by a different particle.

402. (C) The average mass (1×10^{-22} g) ÷ the volume of the nucleus ($^4/_3\ \pi\ (5 \times 10^{-13}\ \text{cm})^3$) $= 2 \times 10^{14}$ g cm^{-3} (π in this equation = 3.14, the ratio of the diameter of a circle to its circumference. It does not refer to the bond). We didn't have to do all that math, however. To estimate an order of magnitude we use $10^{-22} \div (10^{-13})^3 = 10^{-22} \div 10^{-39} = 10^{17}$. That is three orders of magnitude off, which is 1,000 times too big (a big error!), but the AP Chemistry exam won't give us a calculation on the multiple choice section of the exam that requires a calculator to compute in a reasonable amount of time. A question with gigantic or infinitesimal numbers will offer significantly different quantities as answer choices so you can round generously.

403. (C) One half-life = →, so 30 g → 15 g → 7.5 g → 3.75 g → 1.875 g → 0.94 g. The decay to less than 1 gram took five half-lives (five arrows). 325 days ÷ 5 half-lives = 65 days per half-life.

404. (B) In four half-lives, 93.75 percent of a sample decays (see table below). 24 days ÷ 4 half-lives = 6 days per half-life.

Number of half-lives			1		2		3		4		5	
Percent remaining	100	→	50	→	25	→	12.5	→	6.25	→	3.125	
Percent decayed	0	→	50	→	75	→	87.5	→	93.75	→	96.875	

405. (C) After three half-lives, 12.5 percent of a sample remains undecayed (see table in **Answer 404**). If three half-lives take 30 days, each half-life = 10 days.

406. (E) There are six, two-year half-lives in 12 years. The table in **Answer 404** shows that after five half-lives, about 3 percent of the radioisotope remains undecayed (about 1.9 g out of the original 60). One more half-life would reduce that by half, leaving less than, but close to, 1 gram.

407. (B) When an atom's mass number stays the same but the atomic number *increases*, the atom has undergone a β^- (beta) decay. When the mass number of an atom stays the same but the atomic number *decreases*, the atom has undergone a β^+ (positron) decay. About 0.01 percent of K atoms are ^{40}K. β–decay decreases the proton-to-neutron ratio by converting a neutron into a proton and an electron (β-particle, $^{0}_{0}e^-$), the proton stays in the nucleus (so the atomic number increases by one) and the β–particle, the electron, is ejected. Because the *total number* of protons and neutrons stayed the same, the mass number doesn't change.

408. (E) *Radioactive decay displays first-order kinetics.* Choice (E) is a plot of a first-order reaction (with respect to substrate disappearance). Choice (B) is a plot of a second-order reaction (with respect to substrate disappearance).

Chapter 12: Descriptive

409. (D) Silicon combines with oxygen to form a variety of *silicates*. The most common silicate is silicon dioxide, which has a number of different crystalline and amorphous forms. *Silicon dioxide is a covalent network solid*, and the formula *SiO_2 is the empirical formula* of the solid.

410. (C) NaCl is the salt most responsible for the salinity of sea water and the main ingredient in table salt (which may also contain minute amounts of iodine salts, added to prevent iodine deficiency). Any ionic compound with alkali metal is soluble, and compounds formed with alkali and alkali earth metals (groups 1 and 2) as the only metals are generally not colored.

411. (B) $CuSO_4$ is a deliquescent compound. It is white when anhydrous (without water) and turns blue upon hydration. (See **Answer 454** for a summary of properties related to deliquescence.)

412. (D) Most chlorides salts are soluble, with the notable exception of AgCl, $HgCl_2$, and $PbCl_2$. The process of elimination may have allowed us to identify this compound since NH_4NO_3 and NaCl are both white, $KMNO_4$ is purple, and $CuSO_4$ is white or blue depending on its state of hydration.

413. (A) Potassium permanganate, $KMnO_4$ contains manganese. The presence of transition metals in compounds often results in bright colors, even though the pure metals are mostly silver (with the notable exceptions of copper and gold). $KMnO_4$ is a strong oxidizing agent.

414. (B) $CuSO_4$ is a *deliquescent compound*, which means it has a high affinity for water molecules and can absorb them from the air. $CuSO_4$ changes color when it hydrates (it turns blue), making $CuSO_4$ an *excellent desiccant* (because we know when it's full of water and needs to be dried). A desiccant is a substance used to create or sustain a dry environment by absorbing water from it.

415. (E) Ammonium nitrate, NH_4NO_3, is very soluble. (All ionic compounds containing the NH_4^+ ion or the NO_3^- ion are soluble, and this one has both.) It is used for fertilizers not only because it is high in nitrogen, but because the two forms of nitrogen (ammonium and nitrate) are the most readily absorbed and used by plants. The most abundant gas in the atmosphere is N_2 (78 percent). The least abundant is CO_2 (less than 1 percent, but it has considerable effects). Plants can only use the carbon dioxide in the atmosphere as a substrate in organic syntheses, they cannot use N_2 as a source of nitrogen. Only a few kinds of bacteria can metabolize N_2 into a form that is usable by plants.

416. (A) Propane is a gaseous alkane that is often compressed into a liquid for transport. It is a by-product of natural gas processing and the fractional distillation of petroleum. Hydrocarbons are useful fuels because their combustion has a negative free energy change of great magnitude and yields a great number of moles of gas that expand rapidly at the high temperatures produced by the reaction. (They are highly reduced, so there's a lot to oxidize.)

417. (E) Trichlorofluoromethane is a chlorofluorocarbon. It is colorless, practically odorless, and boils at room temperature. Clorofluorocarbons, or CFCs, were used as refrigerants until it became clear that CFCs cause the breakdown of ozone in the presence of ultraviolet light, the frequencies of electromagnetic radiation (10^{15} to 10^{16} Hz, sec^{-1}) from which the ozone layer protects Earth's surface.

418. (C) Hydrogen peroxide is a strong oxidizing agent, which makes it useful for killing bacteria (by oxidative stress). But it also may harm the cells of the body (by oxidative stress), which may prevent healing. Hydrogen peroxide is a natural by-product of all aerobic organisms, but cells contain enzymes that decompose H_2O_2 to O_2 and H_2O at the (low) concentrations produced in cells (as compared to the 3-percent solution in the pharmacy).

419. (D) Hydrogen sulfide has the odor of rotten eggs. It is produced by the decomposition of organic matter under anaerobic conditions and is also found in volcanic gases.

420. (B) Remember that the hydrogen halides *other than HF* (HCl, HBr, and HI) are strong acids. HF is considered weak, but it is still highly corrosive, hence its use in etching glass. Because of its reaction with glass (and metal), HF must be stored in plastic containers.

421. (C) (See **Answer 94** for an explanation of how graphite conducts electricity.)

422. (E) All amino acids contain the carboxyl and amino functional groups. The carboxyl group = COOH (physiological pH, 7.2 – 7.4, is above the pK_a of the carboxyl group, so it is in its *deprotonated* form, COO^-, in body fluids) and the amino group = NH_2 (physiological pH is below the pK_a of the amino group so it is in its protonated form, NH_3^+, in body fluids). The four elements C, H, O, and N make up 96 percent of living matter (easily remembered as CHON).

423. (D) This question is really asking us to recognize *SiO_2 as a covalent network solid* (or at least *not* as a gas) as opposed to asking us to recognize the rest of the answer choices as gases (although it's a good idea to be able to). (See **Answer 11** for an *except* question strategy.)

For **Questions 424–428:**

(A) Isoamyl acetate (banana oil)
(B) Ethyl methyl ether
(C) Benzoic acid
(D) Benzone
(E) Glucose

424. (D) The C=O and the ending "one" make this the ketone.

425. (E) Glucose has five hydroxyl groups per molecule making it very much water soluble.

426. (A) Isoamyl acetate is an ester and is responsible for the fruity odor found in bananas and many other sweet-smelling items.

427. (C) The carboxylic acid group is represented by –COOH.

428. (B) An ether has two alkyl chains attached to a singly bonded oxygen.

Chapter 13: Laboratory Procedure

429. (E) It is not only acceptable but it is *standard practice* to rinse a burette with the solution that will be added to it before it is filled with the solution. This ensures that any water or impurities introduced into the burette during cleaning will be removed. *Any* water or impurities left in the burette will dilute or contaminate the solution, decreasing precision and accuracy. For our safety and the well-being of the balance, *never* place hot objects on a balance. For some substances, this will introduce error into the weighing, as well (air currents are produced around hot objects, and some substances will pick up mass from or lose mass to the environment depending on its temperature).

Adding hot water to a volumetric flask is also considered unsafe, mainly because the mouth of the flask has a small diameter. Using 5 mL of phenolphthalein to titrate 20 mL of acid means that 5 out of 25 mL total volume of our solution is the phenolphthalein indicator. That's 20 percent! Finally, remember the strained but useful rhyme *"Slowly you ought–ta add acids to wat-tah."* The dissociation of an acid (or base) in water is typically exothermic, but for the strong acid and bases, it's *strongly exothermic*. The heat produced by the dissociation can rapidly increase the temperature of the water. Slowly adding acid to water allows the water to absorb excess heat from the dissociation as you pour, so the heat produced can be regulated by the person pouring. It also prevents a harmful splash back. Even if some of the liquid does splash back, it's likely going to be the water that was displaced by the acid that gets splashed instead of the undiluted acid. (See **Answer 430** for a detailed explanation of why we add base to water slowly.)

430. (D) Remember the rhyme, *"Slowly you ought–ta add bases to wat-tah."* The dissociation of strong base in water is *highly* exothermic. Adding the base slowly to water allows the water to absorb excess heat from the dissociation and allows the pourer to adjust the pour accordingly. If the base is added too quickly, or an insufficient amount of water is added to a base, the container can get too hot and can even boil. It can also splatter out of the container due to the vigorous reaction. The composition of the splatter will be a hot (possibly boiling), concentrated solution of a strong base. (See **Answer 429** for a detailed explanation of why we add acids to water slowly.)

431. (D) The best way to deal with an acid spill is to rinse the area to remove excess acid, then neutralize it with a weak base. *Never* use strong acids or bases to neutralize a base or acid spill, especially on skin. Acids and bases neutralize each other in a highly exothermic reaction that can thermally burn skin (in addition to chemically burning it).

432. (B) The most direct and efficient method to determine the *molarity* of this solution is to measure its volume. Two pieces of information are needed to calculate molarity ($M = {}^{mol}/_{L}$), mol and L. The student already knows how many moles of acetic acid are present, so all that needs to be done is to measure the total volume and plug it into the expression for molarity.

433. (B) The most direct and efficient method to determine the *molality* is to measure the mass of the solution. The student needs two pieces of information to calculate molality ($m = {}^{mol}/_{kg\ solvent}$), mol and kg solvent. The student knows the mass of the phosphoric acid, so the number of moles can be calculated. Also, the mass of the phosphoric acid has to be subtracted from the total mass to determine the mass of the solvent, which is needed for the expression.

434. (E) The conductivity of a solution is a measure of its ability to conduct electricity. It can be quantified by measuring the resistance to the flow of electricity between two electrodes in the solution.

435. (B) *Chromatography* is a general term for a laboratory procedure used to separate the components of a mixture. Paper chromatography uses a two-phase system, a solvent and paper, to separate the components of a solution according to their differential solubilities in (or affinities for) the solvent and the paper.

436. (A) The concentration of a colored solution can be determined by *colorimetry*, or visible-light *spectrophotometry*. The more concentrated the solution, the greater the absorbance (or less transmittance) at a particular wavelength. Using standard solutions of known concentrations, a graph of absorbance (or transmittance) versus concentration is constructed so that any absorbance (or transmittance) value can be linked to a concentration by interpolation (obtaining "implied" data from areas on a graph within discrete, measured data points).

437. (D) *Gravimetric analysis* is a method for analyzing the amount of a substance by weighing the products of its reaction with something else that is known. For example, we could determine the concentration of sulfate ions in a solution taking a sample of known volume, adding Ba^{2+} ions until a precipitate stops forming, and then weighing (after filtering and washing) the precipitate. Because we can measure the mass of the precipitate, we know the molar masses of $BaSO_4$ and the stoichiometry of the reaction (Ba^{2+} and SO_4^{2-} react in a 1:1 ratio), and we can figure out how much sulfate was in the original solution. Differential precipitation exploits the different solubilities of ions in the presence of other ions or under different conditions to selectively remove them from a solution. (See **Answer 438** for a description of a similar method of analysis, titration.)

438. (C) *Titration* is a laboratory method that is used to determine an unknown concentration of a known reactant by studying its reaction with a known concentration of a different known reactant. An indicator that the reaction has occurred, or has finished occurring, is required. The two volumes, the concentration of the known solution and the stoichiometry of the reaction, are then used to calculate the unknown concentration. For example, in **Answer 437**, we could have used a Ba^{2+} solution of known concentration to calculate the number of moles of SO_4^{2-} if we had a sensitive way to determine when the very last SO_4^{2-} ions precipitated (an indicator that lets us know we matched up each every SO_4^{2-} ion present with a Ba^{2+} ion). Then we could use $M_1V_1 = M_2V_2$ because we know the molarity of the Ba^{2+} solution (Solution 1), we measured the volume needed to "capture" each SO_4^{2-} (we know they react in a 1:1 ratio), and we know the volume of the SO_4^{2-} solution (Solution 2) we titrated. (See **Answer 437** for a description of a similar method of analysis, gravimetric analysis.)

439. (E) A liquid *in an open container* boils when the vapor pressure above the liquid reaches the pressure atmosphere. Pressure cookers are used to cook faster and at higher temperatures than 100°C. The temperature of water in an open container at sea level cannot exceed 100°C, no matter how much heat is added. *Increasing the pressure in a sealed container is the equivalent of increasing the atmospheric pressure around an open container.* In order to evaporate (and boil), the water molecules must have enough kinetic energy (KE) to reach "escape velocity." The greater the pressure pushing down on the surface of the liquid, the greater velocity (and therefore KE) required to escape. Therefore, a higher temperature (average KE) is required to boil it. Answer choice (A) is not correct because the student would have to add *a lot* of salt to make the water boil significantly higher (the K_b of water is only 0.52 K/m).

440. (E) The volume of most substances changes with temperature, but the mass and the amount of particles don't change with temperature (as long as they are closed systems). Measurements that deal with volume directly (like molarity and density) can be unreliable if the temperature is changed significantly (what constitutes a significant temperature change depends on the substance).

441. (B) A *pipet* will most accurately transfer a particular volume of solution. A flask is not used for the accurate measurement of volumes. A graduated cylinder is appropriate for measuring volumes, but is not the best choice for *the transfer* of a specific volume (too much pouring).

442. (C) The question contains the limiting measurement for significant digits, 50.00 mL.

443. (D) A *volumetric flask* of the proper volume (each flask measures only one volume) is the best piece of equipment for preparing solutions or measuring specific volumes that require a high degree of accuracy. The bottom of the flask is wide and fits most of the sample, but the neck of the volumetric flask is very narrow so the area at the surface of the solution is small. The error incurred by adding an extra mm (in height) of water to a cylinder with an area of 1 cm^2 is insignificant (surface area × length = volume, 0.1 mL in this case) when compared to adding an extra mm of water to a cylinder with an area of 25 cm^2 (2.5 mL) for two vessels of the same volume.

444. (B) NH_4^+ and NO_3^- are both water soluble ions no matter what other ions are present so the only way to retrieve them from an aqueous solution is to evaporate the water and collect the dry solid.

Chapter 14: Data Interpretation

445. (C) Astatine is the heaviest known halogen and is produced by the radioactive decay of other elements. Its short half-life (7–8 hours, depending on the isotope) prevents it from being purified in significant quantities. The astatine purified in the question decayed into bismuth. The bismuth then decayed into lead. We didn't have to know the decay series of astatine to answer this question correctly. Because it has an atomic number of 85, we know it is radioactive and will likely transmutate into other elements.

446. (E) Each compound is only used once, so we can eliminate answer choices even if we're not sure how the particular compound in the question reacts with ammonia.

1 = Silver nitrate
2 = Barium chloride
3 = Copper (II) nitrate
4 = Mercury (I) nitrate
5 = Fe^{3+} ions

447. (A) See **Answer 446**.

448. (C) At low concentrations, NH_3 is a weak base and produces hydroxide ions, OH^-, that combine with Ni^{2+} to form $Ni(OH)_2$, which is normally insoluble. The excess NH_3 allows it to form a stable, soluble, complex ion with Ni^{2+} instead of precipitating with the OH^-.

449. (C) Solid Zn reduces H^+ ions in solution to produce H_2 (Zn is above H in the activity series, and can therefore be oxidized by it). When the CO_3^{2-} ion is present in acidic solutions, CO_2 gas is produced according to this (worth memorizing) reaction:

$$H^+_{(aq)} + CO_3^{2-}{}_{(aq)} \Leftrightarrow H^+_{(aq)} + HCO_3^-{}_{(aq)} \Leftrightarrow H_2CO_3 \Leftrightarrow CO_{2(g)} + H_2O_{(l)}$$

NH_3, however, will not produce a gas when combined with HCl. Instead, the weak base will partially neutralize the strong acid, producing a soluble salt (NH_4Cl) that will remain in solution with the excess H^+ ions.

450. (D) Na_2CO_3 is a very soluble salt (due to the Na^+) and the CO_3^{2-} it produces in solution is a weak base. It will pick up protons to form HCO_3^- and even H_2CO_3. When the CO_3^{2-} ion is present in acidic solutions, CO_2 gas is produced (see the equation in **Answer 449**).

451. (D) The *Law of Multiple Proportions* states that if two elements can combine to form more than one compound, then there is a small, whole-number ratio comparing the masses in which one element combines with the fixed mass of the other. The simplest way to illustrate this is with H_2O and H_2O_2.

	Mass O : Mass H
H_2O	16:2 or 8:1
H_2O_2	32:2 or 16:1

If we hold the mass of hydrogen constant, we can see that the ratio of the mass of oxygen in H_2O_2 compared to water is 16:8 or 2:1. There is no "law of stoichiometry." If there were, it would be a derived from the laws of definite proportions, multiple proportions, and conservation of mass.

452. **(A)** Finding the empirical formula only requires knowing the mole ratio in which the elements in a compound combine.

453. **(C)** Each of the compounds contained 28 g, or ½ mol, Fe. The tricky part is that the masses of oxygen are the masses of molecular oxygen, not atomic oxygen. To find the number of moles of oxygen atoms, we need to find the number of moles of molecular oxygen and double it (there are two oxygen atoms in each molecule). In the first compound, ½ mole Fe combined with ½ mole oxygen atoms (¼ mole O_2 molecules) in a 1:1 mole ratio (FeO). In the second compound, ½ mole Fe combined with ¼ mole oxygen atoms (1/8 mole O_2 molecules) in a 2:1 mole ratio (Fe_2O). In the third compound, ½ mole Fe combined with ~2/3 mole oxygen atoms (~1/3 mole O_2 molecules) in a 3:4 mole ratio (Fe_3O_4).

454. **(D)** *Efflorescence* is the loss of water from a salt crystal upon exposure to air. *Deliquescence* is the property of having a strong affinity to absorb water from the air. It is a property of a good *dessicant*, a substance used to create and/or maintain a state of dryness. *Hydrophilic* is synonymous with the term *water-soluble*. (See **Answer 11** for an *except* question strategy.)

455. **(B)** Blue, or dry, $CoCl_2$ has a molar mass of 130 g mol^{-1}, so 13 g = 0.1 mole. The sample absorbed 4 g of water from the atmosphere during handling, which is equal to ~0.2 mole of water (twice the number of moles of $CoCl_2$). The formula for the purple hydrate is thus $CoCl_2 \cdot 2\,H_2O$. The sample absorbed 7 additional g of water, for a total of 11 g, or 0.6 mole of water, making the formula for the red hydrate $CoCl_2 \cdot 6\,H_2O$.

456. **(B)** With the exception of gold and copper, the color of all of the pure transition metals (as reduced metals) is silver. *The compounds of transition metals are colored*, and these colors of are due to electron transitions that occur when the metal bonds with other elements. The oxidation state of the transition metal determines the color of the complex. The two most common types of electron transitions are *electron transfers* and *d-d transitions.*

Electron transfer: In a complex ion, ligands (ions or molecules that bond with the central metal) surround the metal and form a coordination complex. Often, the ligand transfers one or more of its electrons to the central metal atom (they are commonly thought of as Lewis bases). In complexes in which the central atom is a transition metal, the *electrons* involved in the *transfer* are easily excited by wavelengths in the visible light spectrum, causing some wavelengths of light to be absorbed and others to be reflected. The reflected wavelengths are those we see as colors.

d–d transitions: In transition metal complexes, the d orbitals vary in energy level. These differences in energy correspond to the wavelengths of light that can be absorbed (and reflected).

457. (B) The molar mass of a compound can be found by its vapor density. The formula is derived from the ideal gas law: PV = nRT but substitute $(^{\text{mass,m, in g}}/_{\text{MM (molar mass)}})$ for n (the number of moles) and solve for molar mass (MM).

$$MM = {}^{g\,R\,T}/_{P\,V} \therefore MM = {}^{(3.0)\,(0.0812)\,(373)}/_{(1)\,(2)} \therefore MM = 46 \text{ g mol}^{-1}$$

458. (C) If the temperature was not actually brought up to 100°C but the calculation was still done using the value 373 K, the molar mass calculated would have been *larger* (T is in the numerator in the equation derived in **Answer 457**). (See **Answer 11** for an *except* question strategy.)

459. (B) The molar mass (MM) of a compound can be found by measuring its effect on the freezing point or boiling point of a liquid. In this case, we know that 10 g of solute added to 50 grams of water raised the boiling point by 2°C.

The formula we use to determine freezing point depression or boiling point elevation is $\Delta T = K_b mi$, where K_b is the ebullioscopic constant, a constant that allows us to relate the molality of a solution to the boiling or freezing point of the solvent (subscript b is for boiling point). The K_b and K_f are specific to a particular solvent. The van't Hoff factor, i, is the ratio of the number of moles of particles a compound produces in solution relative to the number of moles of particles of compound added.

The formula $\Delta T = K_b mi$ can be used to determine the numerical value of m (molality = $^{\text{mol solute}}/_{\text{kg solvent}}$). The mass of the solute divided by its molar mass (MM) gives us the number of moles of solute; while the number of moles of solute divided by the mass of the solvent gives us the molality. Therefore, we can rewrite the m expression as $m = (^{\text{mass of solute}}/_{\text{(MM)(kg solvent)}})$. We know the mass of the solute and we calculated m with the equation $\Delta T = K_b mi$, so now we rearrange our molality expression to solve for MM: MM = $^{\text{mass of solute}}/_{\text{(kg solvent)}(m)}$.

Alternatively, we could substitute the expression for m, $(^{\text{mass of solute in g, grams}}/_{\text{(MM)(kg solvent)}})$, directly into the equation $\Delta T = K_b mi$ and solve for MM:

$$MM = {}^{i\,K_b\,g}/_{kg\,\Delta T} \therefore MM = {}^{(2)\,(0.51)\,(10)}/_{(0.05)\,(2)} = \sim 101 \text{ g mol}^{-1}$$

460. (C) The point in the curve where the temperature (average kinetic energy) is stable means the potential energy of the substance is changing (decreasing in this case). The absence of a temperature change in a substance that is gaining or losing energy is an indication that the substance is undergoing a phase change.

461. (C) The temperature of the solution increases about 1°C and then decreases again to 70°C over the course of about three minutes. The average slope of the line during that time period is close to zero, and indicates the approximate freezing point of the solution.

462. (B) Use $\Delta T = K_f mi$, i = 1 (no dissociation).

$$\Delta T = 10°C \therefore 10 = (7.1)m \therefore m = \sim 1.4$$

463. **(B)** $\Delta T = K_f mi$, i = 1 (no dissociation)

$\Delta T = 12.5°C, K_f = 5 \therefore m = {}^{\Delta T}/_{K_f} = {}^{12.5}/_{5} = 2.5m$

$\text{molality} = {}^{\text{mol solute}}/_{\text{kg solvent}} \therefore \text{moles solute} = (m) \times (\text{kg solvent})$

moles solute = (2.5)(0.10) = 0.25 mol solute

moles solute = ${}^{\text{mass solute}}/_{\text{molar mass}}$, solving for molar mass:

$\text{molar mass} = {}^{\text{mass solute}}/_{\text{mol solute}} = {}^{20\text{ g}}/_{0.25\text{ mol}} = 80\text{ g mol}^{-1}$

464. **(C)** Diffraction is the spreading out of waves when they're passed through small openings. Diffraction is a property of waves, and to occur, the wavelengths of the waves must be comparable in length to the size of the opening through which they are being passed. The pattern that emerges is similar (and probably indistinguishable from) an interference pattern, with alternating dark and light bands. X-rays have wavelengths comparable to the spaces between the ions in an ionic crystal lattice and will produce a diffraction pattern which is used to analyze the lattice structure of the crystal.

465. **(E)** The calculation for determining the percent mass of a hydrate =

$({}^{\text{mass of dry sample}}/_{\text{mass of hydrated sample}}) \times 100$. If the sample was calculated to have only 26 percent water, it's because the mass of the dry sample was greater than its actual mass, making the dry compound a larger percent of the hydrate than it actually is. Deliquescent compounds readily absorb water from the atmosphere, so the most likely cause of the increased mass of the dry compound is that it wasn't really dry; it still had water molecules attached to it when it was weighed.

466. **(A)** Mg is above hydrogen in the activity series, so we know that it can be oxidized by the H^+ ions in the HCl solution. The Mg will reduce 2 H^+ to H_2 gas. Zn is also above hydrogen in the activity series, but the Zn in $Zn(NO_3)_2$ has already been oxidized (it's in the form of Zn^{2+}), and so it has no electrons with which to reduce H^+.

Na^+ in its elemental form could certainly reduce H^+, but not in its oxidized form. The CO_3^{2-} will produce CO_2 gas in an acidic solution, not H_2 gas, according to the equation

$$H^+_{(aq)} + CO_3^{2-}{}_{(aq)} \Leftrightarrow H^+_{(aq)} + HCO_3^-{}_{(aq)} \Leftrightarrow H_2CO_3 \Leftrightarrow \mathbf{CO_2}_{(g)} + H_2O_{(l)}.$$

467. **(C)** We are titrating a weak base with a strong acid. HCl is a strong acid, but since the pH at the equivalence point is < 7, we know we are titrating a weak base (the pH of the equivalence point of a strong acid-base titration = 7). *Methyl red* is the best indicator listed for this titration because it changes color with the pH range of the equivalence point of the titration (the equivalence point pH is it about 5 and methyl red undergoes a color change between pH 4.4 and 6.2).

468. **(A)** A buffer solution resists changes in pH by acting like a "proton sponge," giving them to the solution when the $[H^+]$ decreases and absorbing them from the solution when the $[H^+]$ increases. The buffer region in this titration consists of the weak base and its conjugate acid. The buffer region is always at the beginning of the titration, when the weak base and its conjugate acid are present in fairly high concentrations, before too much neutralization has occurred. In this titration, the buffer region lies between pH 11 and 8. *Phenolphthalein* would remain pink in the buffer region and transition to clear at the end of it.

469. (B) The CO_3^{2-} ion would have produced CO_2 gas in an acidic solution (Sample 1) according to the equation

$$H^+ + CO_3^{2-} \Leftrightarrow H^+ + HCO_3^- \Leftrightarrow H_2CO_3 \Leftrightarrow \mathbf{CO_{2(g)}} + H_2O.$$

$CaSO_4$ and $BaSO_4$ both form an insoluble white precipitate, while aqueous NH_3 precipitates green $Ni(OH)_2$ and forms a colored solution when it reacts with water and ammonia according to the equation

$$Ni^{2+} + 2\,NH_3 + 2\,H_2O \Leftrightarrow Ni(OH)_2 + 2\,NH_4^+.$$

470. (D) The purpose of collecting a gas over water is to keep the gas at a constant pressure as more gas particles are collected. Gases are highly compressible, so their volume changes with changes in pressure. Measuring the amount of gas in a closed, rigid container requires that *both* pressure and volume be measured.

An eudiometer, a device used to collect gas over water, works by displacing water with gas, but the pressure of the gas always equilibrates to the atmospheric pressure because the gas pushing down on the surface of the water in the eudiometer is the same (at equilibrium) as the pressure of the atmosphere pushing the water up in the eudiometer (by pushing down on the surface of the water in the container in which the eudiometer is placed).

471. (D) HCl and NH_3 are *very* water soluble gases. A large number of the gases will dissolve in the water. They must pass through to be collected, greatly reducing the yield. CO_2 is slightly soluble, so it's not the best gas to collect over water, but a relatively small fraction will be lost to its dissolving in (and reacting with) water compared to HCl and NH_3.

472. (B) The gas collected in the eudiometer is actually a mixture of two gases—the gas intentionally collected and water vapor that evaporated in the eudiometer. The partial pressure of the water vapor is determined solely by the temperature. At 22°C, the temperature at which the experiment was performed, the vapor pressure of water is 19.8 torr. The total pressure of the gases is 770 torr (the atmospheric pressure). Dalton's law of partial pressures states that the total pressure of a mixture of gases is the sum of the partial pressures of each of the gases in the mixture $\therefore$ 770 − 19.8 = 750.2 torr. (See **Answer 470** for an explanation of why the total pressure of the gases in the eudiometer is equal to the atmospheric pressure.)

473. (B) $\text{mass}_{\text{cup+water}} - \text{mass}_{\text{cup}} = \text{mass}_{\text{water}}$

$$\text{mass}_{\text{cup+water+ice}} - \text{mass}_{\text{cup+water}} = \text{mass}_{\text{ice}}$$

474. (E) The mass of the ice *measured* would have been greater than the actual mass of the ice, so the heat of fusion calculated would be too low. The units of $H_{\text{fusion}} = {}^{\text{kJ}}/_{\text{gram}}$ or ${}^{\text{kJ}}/_{\text{mol}}$. Either way, a larger mass would have increased the value of the denominator and decreased the value of H_{fusion}.

475. (C) To calculate the heat of fusion of ice using this method, the specific heat of the ice must be known since the temperature of the ice must be raised from −20 to 0°C before it melts. In addition, the heat used to raise the temperature of the ice and to melt it comes from the *loss of heat* of the liquid water, so the specific heat of liquid water must also be known. H_{fus} = (total heat lost by liquid water) − (heat gained by ice to reach 0°C).

476. (B) 0.4 M = 0.4 mole per liter, or $^{0.4\ mol}/_{L}$ solution × 0.250 L = 0.1 mole KOH (since the L cancels). Remember to convert to L; *do not use mL.* The molar mass of KOH = 56 g mol^{-1} ∴ 0.1 mole = 5.6 g.

477. (D) This question refers to an electrolytic cell. Two hours reduced 230 grams (10 mol) of Na^+ into Na. Since 1 mole of electrons is needed to reduce 1 mole of Na^+ ions, *10 moles of electrons must have passed into the cell in two hours.* Fe^{3+} requires *3 moles of electrons per mole,* so we intuitively know we'll get a little more than one-third the number of moles of Fe metal. 10 moles electrons × ($^{1\ mol\ Fe^{3+}}/_{3\ mol\ electrons}$) = 3.3 moles Fe^{3+} ions can be reduced. The molar mass of Fe is 56 g mol^{-1} ∴ 3.3 moles Fe weighs about 185 g.

478. (D) We don't really need to calculate the final concentrations of ions in the resulting solution because they are all present in the same volume, so the total number of moles of each species is enough to compare their relative concentrations.

0.1 L of $^{2\ mol}/_{L}$ $Pb(NO_3)_2$ = 0.2 mole $Pb(NO_3)_2$, which yields 0.2 mole Pb^{2+} ions and

2 × 0.2 = 0.4 mole NO_3^- ions.

0.1 L of $^{3\ mol}/_{L}$ NaCl = 0.3 mole NaCl, which yields 0.3 mole Na^+ and 0.3 mole Cl^- ions.

$PbCl_2$ will precipitate out of solution in a 1:2 ratio of Pb^{2+} to Cl^-. 0.2 mol Pb^{2+} will precipitate 0.4 mole Cl^- ions, but there are only 0.3 mole, so just about all the Cl^- ions will be taken out of solution by the 0.15 mole Pb^{2+} ions (0.3 mole Cl^- will combine with 0.15 mole Pb^{2+} leaving 0.2 − 0.15 = 0.05 mole Pb^{2+} behind).

479. (D) Differential precipitation can separate ions mixed in a solution.

	Ba^{2+}	Fe^{3+}	Zn^{2+}
Cl^-	soluble	soluble	soluble
OH^-	soluble	INSOLUBLE	INSOLUBLE
SO_4^{2-}	INSOLUBLE	soluble	soluble

480. (A) An electron gets excited (raised to an orbital of higher energy) when its atom or molecule *absorbs* energy. The excited electrons are not stable at the higher energy levels, and when they drop back down to ground state, they emit energy. That energy is often lost as a photon (but there are other ways for energy to be lost). *The energy of the photon is proportional to the energy change of the electron.*

481. (D) The photoelectric effect demonstrated the particle-like properties of light (or more generally, electromagnetic radiation).

482. (D) $E = h\nu \therefore E = (6.63 \times 10^{-34})(4.4 \times 10^{14}) = 2.9 \times 10^{-19}$ J.

483. (B) The energy of a photon is given by the equation $E = h\nu$, where $h = 6.63 \times 10^{-34}$ J·s and ν = the frequency. Make sure to use the frequency and *not* the wavelength.

484. (C) Each line in line spectra corresponds to an energy change of an electron. Any one electron can produce several lines according to the energy transitions it experiences. (See **Answer 11** for an *except* question strategy.)

485. (C) Waves can be imagined as vibrations produced by an oscillator. When the waves from an oscillator are polarized, all the vibrations occur in one plane.

486. (D) The spin of an electron is either +½ or −½ , so only two spots would be expected to appear on the screen. The electrons spinning with +½ spin would be deflected in one direction by the field, and the electrons with the −½ spin would be deflected in the opposite direction.

487. (E) Emission lines represent the energy transitions of the electrons in an atom when energized. A magnetic field applied to the gas will *not change* the magnitude of the energy transitions of electrons occupying the same orbitals, but applying the field will cause the electrons to behave as magnets within the field. Electrons of opposite spins are like opposing poles of a magnet. The field will affect them in opposite directions, splitting the emission lines they produce.

488. (B) See **Answer 486**.

489. (C) The photoelectric effect is evidence that light (or more generally, electromagnetic radiation) has particle-like properties. When photons of a high enough frequency strike a metal plate, electrons are dislodged from the plate. Waves do not have this property. All photons are massless, they travel at the same speed in the same medium, and their energy is determined *only* by their frequency.

490. (E) When J.J. Thomson performed his cathode ray experiment to determine the charge-to-mass ratio of electrons, he knew the cathode ray was composed of electrons. Their deflection in magnetic and electric fields indicated they were charged. A variation of Young's *double-slit experiment* using electrons instead of photons showed that electrons have the wave-like property of interference.

491. (B) Cancelling out units is the easiest way to solve this problem. We need to combine C and $^{C}/_{g}$ to get g, so we can flip $^{C}/_{g}$ to $^{g}/_{C}$ and the C unit cancel:

$$-1.602 \times 10^{-19}\ \text{C} \times (^{1\ \text{gram}}/_{-1.76 \times 108\ \text{C}}) = (1.602 \times 10^{-19}\ \text{C}) \times (-1.76 \times 10^{8})^{-1}.$$

492. (D) Neutrons are the only particles in the list that have no charge. Any charged particle will be deflected when passed through an electric or magnetic field.

493. (B) Liquids with strong intermolecular forces have a high surface tension due to the strong attraction of surface molecules to each other. Strong forces of attraction between the molecules in a liquid cause the surface molecules to stick to each tightly, creating the appearance of a film. The greater the tension of the film, the steeper the meniscus it creates.

494. (E) *Viscosity* is defined as a fluid's resistance to flow. Liquids with strong intermolecular forces will have a high viscosity because the particles in the liquid resist moving past one another.

495. (E) See **Answer 494**.

496. (E) *Miscible* typically refers to two or more liquids that are soluble in each other. Miscible sounds like mixable, and they can be thought of as synonyms. Water and glycerol can both form hydrogen bonds (The –ol ending in glycerol tells us it's an alcohol and ∴ has O—H bonds. Glycerol is actually a three-carbon compound with a hydroxyl group on each carbon, for a total of three.). The high surface-tensions of water and glycerol and the high viscosity of glycerol indicate that their particles experience strong intermolecular forces of attraction.

497. (C) A *Newton* (N) is a unit of force. Surface tension is measured by the force required to break the surface of a liquid. The surface tension of castor oil is expected to be lower than water and glycerol since the only intermolecular forces of attraction in oils are weak London dispersion forces. We would expect the surface tension to be similar to that of olive oil but larger than benzene (castor oil and olive oil are triglycerides, or triacylglycerols), which have very high molar masses (in several hundred g mol^{-1}).

498. (B) The breaking of van der Waals forces (intermolecular forces of attraction, IMFs) is endothermic. Higher temperatures disrupt IMFs, decreasing surface tension and viscosity.

499. (B) A very small fraction of the molecules of a weak acid dissociate in solution, so we expect slightly more than 1 M of particles (but much less than two).

500. (E) A good electrolyte produces a high concentration of ions in solution. Compound E produces three moles of ions per mol compound when dissolved. Soluble salts and strong acids and bases are excellent electrolytes because they completely dissociate in solution, producing at least two moles of ions per mole of compound.